T0271832

Permeation Grouting for Liquefaction Countermeasures

Academic and industry experts describe the use of chemical (permeation) grouting beneath an airport runway to improve ground resistance to liquefaction. They present the cost, environmental, and operational benefits; specifications; methodology; and practical results of this cutting-edge method. Because transportation infrastructure such as ports and airports are required to operate even in the event of a large earthquake, they must be resilient against liquefaction.

Through contributions from experts in academia and industry, this book describes the discovery of construction defects at three Japanese airports and the subsequent project to repair and strengthen the ground using chemical grouting with environmentally friendly colloidal silica, the first time this technique has been used in Japan. This book first describes chemical grouting and its benefits, its specifications, and field investigation results of its ground improvement performance. Next, it demonstrates a numerical and probabilistic method to model spatial variability in material properties of field data on improved ground. Finally, it explains a performance-based verification for airport runway availability in terms of bearing capacity and runway flatness after a large earthquake. Through its clear explanations, this book enables readers to implement chemical grouting and enjoy the cost, environmental, access, and operational benefits of this technique over traditional methodologies that would require temporary site closure and large-scale excavation.

Because the concept and methodology described in this book are applicable to various geological, geotechnical, and seismological conditions depending on the location and structural and operational conditions contingent on the infrastructure type, this book is a useful resource for geotechnical and other infrastructure engineers who must strengthen the ground without disrupting normal operations.

Permeation Grouting for Liquefaction Countermeasures

Implementation and Performance Evaluation

Edited by
Kiyonobu Kasama
Yoshihisa Sugimura

CRC Press is an imprint of the
Taylor & Francis Group, an **informa** business

Designed cover image: Shutterstock

First edition published 2025
by CRC Press
2385 NW Executive Center Drive, Suite 320, Boca Raton FL 33431

and by CRC Press
4 Park Square, Milton Park, Abingdon, Oxon, OX14 4RN

CRC Press is an imprint of Taylor & Francis Group, LLC

ISBN: 978-1-032-67011-9 (hbk)
ISBN: 978-1-032-67012-6 (pbk)
ISBN: 978-1-032-67013-3 (ebk)

DOI: 10.1201/9781032670133

Typeset in Minion
by SPi Technologies India Pvt Ltd (Straive)

Contents

Preface

JAPAN IS AN EARTHQUAKE-PRONE COUNTRY, AND LIQUEFACTION OF GROUND has caused extensive damage to social infrastructure. Transportation infrastructures such as ports and airports are required to maintain their operations even after a large earthquake, because they act as transportation hubs following a disaster. From these backgrounds, Japan has developed multiple ground improvement technologies as a liquefaction countermeasure. Chemical grouting method is a new ground improvement technology that can be applied below existing structures such as airport runways without removing the existing structures.

In 2016 construction defects were discovered at three Japanese airports where liquefaction countermeasure construction was initiated to improve seismic performance. In each case, because construction work would be conducted while the airport runway remained in service, a special chemical grouting method was adopted. The chemicals, however, were not injected properly, which led to the world's first attempt to apply liquefaction countermeasures to heterogeneous ground that had not been sufficiently injected with chemicals. The contents presented in this book are geotechnical engineering findings from a repair project for defective liquefaction countermeasure construction at an airport in earthquake-prone Japan.

This book introduces the chemical grouting method and its ground improvement performance based on the field investigation. It also proposes a method to evaluate the effect of ground improvement based on "performance-based specifications" rather than conventional "prescription-based specifications" by applying the reliability analysis for improved ground against liquefaction that was developed and deployed by the authors over the past several years.

In other words, the bearing capacity and deformation characteristics of the airport runway, which is improved by the chemical grouting method as a liquefaction countermeasure due to partial ground liquefaction, are probabilistically and statistically investigated by modeling the improved ground with a finite element analysis considering shear strength autocorrelation and expressing strength spatial variability in random field theory.

The authors introduce "the percentage of defective for ground improvement," which is the volumetric ratio of on-site shear strength that does not satisfy the design undrained shear strength value. The authors have also developed a chart showing the relationship among the percentage of defective for ground improvement, the bearing capacity and airport runway flatness after partial liquefaction. The charts are used to evaluate the effectiveness of ground improvement against liquefaction and determine the performance-based reliability of the airport runway resilience against earthquakes. These charts are equivalent to the highest level of Level 3 reliability design and highly practical, as they enable simple verification of bearing capacity and deformation of the liquefied ground based on "performance-based specifications."

The method presented in this book has the potential to be widely applied not only to the evaluation of chemical grouting improved ground but also to the evaluation of various soils with spatial variations in soil properties based on "performance-based specifications." This is an extremely significant feature that promotes the sound development of geotechnical engineering.

Kiyonobu Kasama

About the Editors

Kiyonobu Kasama is a Professor of Geotechnical, Geoenviromental Engineering and Geodisaster Prevention Laboratory at Kyushu University. He obtained his Dr. Eng from Kyushu University, where he has been teaching and conducting research for more than 20 years. He received the Outstanding Paper Awards for Young Researchers from the Japanese Geotechnical Society and The Society of Material Science, Japan.

Yoshihisa Sugimura is a Professor at the Graduate School of Maritime Sciences, Kobe University. After receiving his MSc from The University of Tokyo, he joined the Ministry of Land, Infrastructure, Transport and Tourism in 1999, a position that included a secondment to the Ministry of the Environment. He received his Ph.D. degree from The University of Tokyo.

CHAPTER 1

Background of This Book

Yoshihisa Sugimura

1.1 INTRODUCTION

Japan is an earthquake-prone country, and tremendous damage due to liquefaction from earthquakes has been documented. The social infrastructure that safeguards people's lives and the economy must appropriately cope with seismic risk, and airports—which are transportation infrastructure—must continue to operate even in the event of a massive earthquake, as they serve as transportation hubs following disasters.

In 2008, the Japanese government issued a document, "Resilient Airports Against Earthquakes," which outlines the roles required for airports in the event of earthquakes. According to the document, "airports that serve as hubs for emergency transportation" are required to function as hubs for emergency and lifesaving activities immediately following a disaster, and must be able to receive emergency supplies and personnel within three days after a disaster. In addition, "airports important for air transportation" are required to facilitate regular commercial aircraft operations within three days of the disaster, a transportation function equivalent to 50% of normal operations as soon as possible after reopening to halve the economic damage caused by the earthquake, as well as maintenance of the aviation network, sustainability of economic activities in the hinterland, and maintenance of the metropolitan area's functions.

DOI: 10.1201/9781032670133-1

In response to this, airport disaster prevention base plans including required functions have been developed at each airport, and earthquake resistance measures such as liquefaction countermeasures have been implemented sequentially. In the Great East Japan Earthquake of 2011, which caused enormous damage mainly in the Tohoku region of Japan, liquefaction countermeasures that had already been implemented at Sendai Airport near the epicenter of the earthquake, proved to be effective. While the areas where liquefaction countermeasures had not been implemented were severely damaged, the runway at Sendai Airport maintained its functionality after the earthquake and was ready for use immediately. The importance of earthquake resistance measures is evidenced by such cases.

However, in May 2016, construction defects were detected in five ground improvement projects at three airports that had been implemented as earthquake resistance measures. In some construction projects using the chemical grouting method, it was discovered that they had been completed without injecting an sufficient amount of chemical to the design value. The percentage of chemicals injected in the five projects with construction defects ranged from 5.4 to 76.8%, and the creation rate of improved bodies with sufficient improved diameters as planned ranged from 0 to 23.3%. Although it is naturally a social problem that the earthquake resistance of the airports was not ensured and that the required functions could not be performed during an earthquake, the repair work for this kind of construction defect was an unprecedented project in which the half-finished ground had to be re-modified, and the project faced the geotechnical challenge of the large variation in ground conditions. Therefore, the key point was how to evaluate the ground, and it was considered to be more appropriate to evaluate the effect of ground improvement in terms of performance than to strictly implement the specifications (such as the amount of chemical injection) set for the repair work.

In this regard, prior to construction, detailed ground investigation and test construction were conducted, and emphasis was placed on the appropriate evaluation of the strength of the existing and improved ground, and the certainty of construction quality, such as ground improvement effects. The results of the performance verification confirmed that the shape, quality, and performance were secured in all areas where chemical injection had been completed, and all repair works were completed in December 2020.

This book presents new geotechnical findings from the repair work of one of the airports where construction defects occurred.

1.2 OUTLINE OF CONSTRUCTION DEFECTS

When ground improvement is performed immediately under an existing structure using chemical grouting, a drilling method in which the boring track is bent under the structure while drilling from a location where there is no structure (bend drilling) is used because it is difficult to perform the usual vertical drilling method. When ground improvement is performed immediately beneath a runway or taxiway at airports in service, it can only be done during limited nighttime hours if construction is performed from atop the runway or other structures. For this reason, a bend hole drilling method is utilized to ensure sufficient construction time, which allows drilling from an open space beside the runway to immediately beneath the runway. Four of the five projects found to be defective utilized bent drilling, but the accuracy of the drilling hole position was low, ranging from 0 to 54.8%, and this was one of the factors causing the construction defects.

Another factor was a problem related to chemical injection. In the five construction projects where construction defects were found, it was specified that the ground improvement work would be performed by chemical injection, but it was not specified which of the chemical grouting methods was to be used. As a result, a relatively new method was adopted, and there were no actual applications of this method, especially with long-distance bend drilling, resulting in the application of an immature method to the sites.

In fact, backflow and leakage of the chemical were observed relatively early after the start of injection, and attempts to deal with these problems were ineffective due to the strict control of the impact on runways and other facilities at airports in service. As a result, insufficient chemical was injected to the design value, and even when some amount of chemical was injected, the production rate of improved bodies with sufficient improvement diameter as planned was extremely low. Figure 1.1 shows an image of the above construction defects, and Table 1.1 provides a summary (Expert Committee on Problems Related to Ground Improvement Construction Defects, 2016).

Among the construction works shown in Table 1.1, the geotechnical knowledge obtained from the repair work for defective construction at X Airport, which is marked, is the subject of this book.

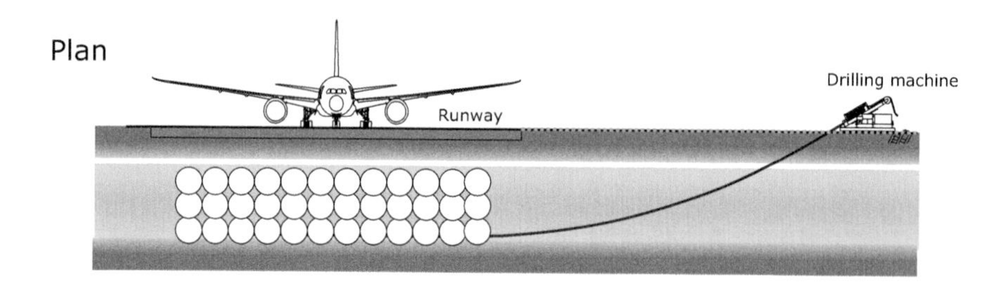

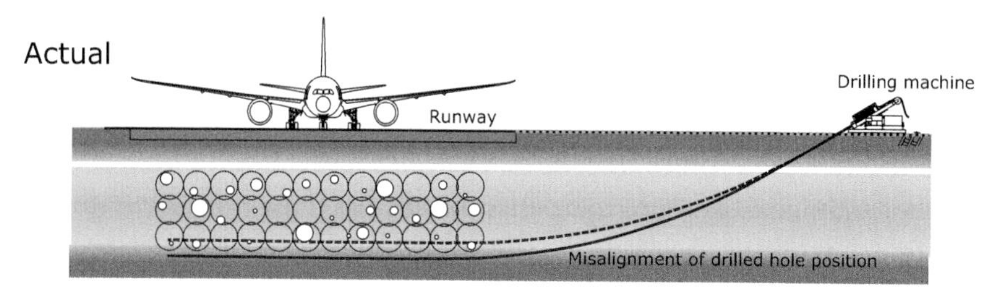

FIGURE 1.1 Image of construction defects.

TABLE 1.1 Summary of Defective Construction Projects

Airports/Facilities	**Runway of X Airport**		**Runway of Y Airport**	**Taxiway of Y Airport**	**Taxiway of Z Airport**	
Construction	A	B	C	D	E	
					Side of underpass	**Underneath of underpass**
Accuracy of drilling hole position	39.9%	54.8%	0%	—	—	0%
Percentage of chemical injection	42.8%	37.7%	5.4%	45.0%	48.4%	76.8%
Creation rate of improved bodies	12.9%	1.4%	0.0%	5.7%	23.3%	0.0%
Subject of this book	✓	✓				

1.3 REPAIR OF GROUND IMPROVEMENT PROJECTS

The repair of the construction defects was an unprecedented project to re-modify the half-improved ground and was highly challenging from a geotechnical engineering standpoint, as the variability of the ground conditions was a major issue. Therefore, it was decided to establish a committee of experts for each airport to conduct a technical study through in-depth ground investigation and test construction, and to decide on the method of repair work. The following describes the process of examining the repair of the X airport that is the subject of this book.

The post-construction ground at the X airport is defective constructed ground in which approximately 40% of the design volume of the chemical was injected heterogeneously. Since the variability of ground conditions is a serious issue in re-improving this ground, it is important to evaluate the entire defective ground. As a clue to this, it was considered that if the relationship between injection results and strength and the relationship between soil properties and strength could be clarified, it would lead to the evaluation of defective ground by utilizing the actual injection results at each known injection point from actual construction data. The strength confirmation by sampling was adopted in the actual ground investigation because the relationship with strength is uncertain except for sampling as a ground evaluation method.

As a result of the ground investigation, it was confirmed that the hydraulic conductivity of the defective ground was lower than that of the ground before the construction work (estimated value) due to the effect of the chemical solution, that the fine grain content was less than 40%, and that the unconfined compression strength of the "inside" of the assumed improved body was greater than that of the "outside" as a result of the location relationship between the assumed improved body and the sampling points. Considering that the results of preliminary tests also confirmed the applicability to low permeability ground, it was judged that the ground was suitable for the permeation grouting method, which was adopted as the repair procedure.

The repair work was unprecedented among the similar projects, and the important point is that the following actions were adopted (Kyushu Regional Development Bureau, Ministry of Land, Infrastructure, Transport and Tourism, 2022).

- After confirming the ground conditions (effects of prior work) through on-site ground investigation, conducting "preliminary experiments" on ground that simulates construction defects, and conducting "on-site test construction" at a part of the repair site to confirm applicability, the ground improvement method (permeation grouting method) to be used for the repair work is selected.

- New quality evaluation methods were introduced, and a comprehensive geotechnical evaluation with performance verification was conducted to properly evaluate the construction results (shape and quality).

1.4 CONCLUSIONS

The contents presented in this book are geotechnical engineering findings from a repair project for defective liquefaction countermeasure construction at an airport in earthquake-prone Japan. This was the world's first challenge to apply liquefaction countermeasures to heterogeneous ground that had not been sufficiently injected with chemicals. Therefore, the findings are diverse and organically related to each other. Specifically, findings include the chemical grouting method and its ground improvement performance based on the field investigation, a numerical and probabilistic method to model spatial variability in material properties of field data on improved ground, and a performance-based verification for airport runway availability in terms of bearing capacity and runway flatness following a massive earthquake. In other words, this is the first book on the practical application of performance-based verification to an existing airport.

REFERENCES

Expert Committee on Problems Related to Ground Improvement Construction Defects, 2016. "Interim report of Expert committee on problems related to ground improvement construction defects." https://www.mlit.go.jp/common/001140791.pdf

Kyushu Regional Development Bureau, Ministry of Land, Infrastructure, Transport and Tourism, 2022. "Material for the 10th Expert committee on the repair of ground improvement works at Fukuoka airport." https://www.pa.qsr.mlit.go.jp/iinkai/pdf/2020122403.pdf

CHAPTER 2

Chemical Grouting Method

Yoshiyuki Morikawa and Shinji Sassa

2.1 INTRODUCTION

Chemical grouting is a ground improvement method in which chemicals, whose hardening time can be arbitrarily adjusted, are injected through an injection pipe installed in the ground to increase imperviousness or strength of the ground. Chemicals are injected into voids among soil particles without changing the structures of soil skeleton and solidify after hardening time. Injection of chemicals causes no damage to existing structures. The chemical grouting method is, therefore, often employed for ground improvement, especially to prevent liquefaction, of ports and airports in service. This chapter provides fundamental information, such as preliminary survey, design (setting of the specifications of grout and mix proportion design of chemicals), and implementation for this method.

2.2 OUTLINE OF CHEMICAL GROUTING

2.2.1 Mechanism of Chemical Grouting

The chemical grouting method is mainly employed to improve sandy ground. Grouting does not mean filling cavities behind structures and closing joints or cracks of concrete. In this method, chemicals replace pore water by injection into voids among soil particles and solidify after designed hardening time. Improvement by this method increases cohesion but no increase of internal friction, because injection is carried out without

DOI: 10.1201/9781032670133-2

change and collapse of the structures of the soil skeleton. Furthermore, it is possible to drastically reduce permeability of the ground since pore water is replaced with solidified chemicals.

2.2.2 Chemicals

The materials injected into the ground include non-chemicals such as cement and clay, and chemicals such as water glass and polymers. The use of polymer in Japan has been prohibited to be applied except for emergency use for face stability of mountain tunnels. Therefore, the injected material is limited to cement as non-chemicals and water glass as chemicals. It is difficult to control stabilization time of non-chemicals, of which the stabilization mechanism is based on the hydration process, whereas that of chemicals can be controlled by adjusting mix proportion of materials of chemicals. Injection of non-chemicals, therefore, cannot be based on the concept and specifications of chemicals, provided in this book.

Chemicals can be classified into slurry type and solution type depending on the presence of suspended particles of binder, such as cement and slug. Cement-Bentonite consisting of cement and clay, for example, is categorized as slurry type, and active silica produced from water glass as solution type. The slurry type chemicals are superior to the solution type in achieving strength of improved body, while below in size of the range of uniform grouting. Figure 2.1 shows the classification of chemicals for Grouting.

Chemicals are produced by mixing solutions of water glass and hardening agent that are refined separately and injected into the ground. Solutions harden after mixing, with an increase in viscosity, and then lose fluidity at the designated position in the ground. The phenomenon of losing fluidity is called gelation. The time from when solutions are mixed to when the chemicals lose fluidity is defined as "gelling time."

2.2.3 Safety of Chemicals

Water glass, which is an inorganic chemical material used for various purposes, is the main material of chemicals in quantity. The safety of water glass is so high that it can be used for purification of tap water and as an additive in soap, etc. (Matsuoka, 2021). Hardening agent is used for solidification of water glass, and its amount in chemicals is much smaller than that of water glass; hardening agent is also safer than water glass. The chemicals are injected into the ground as a mixture of water glass and hardening agent. These materials are thoroughly mixed and solidified,

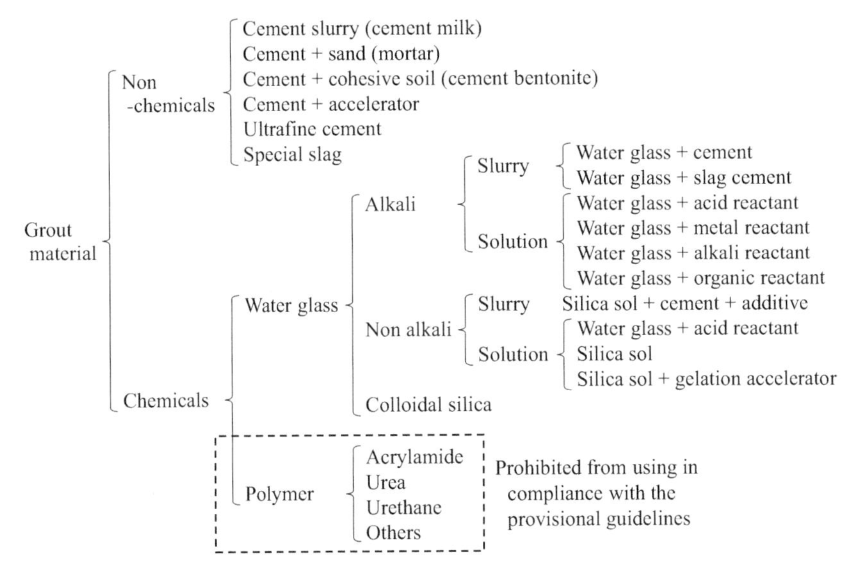

FIGURE 2.1 Classification of chemicals for grouting. (After The Overseas Coastal Area Development Institute of Japan, 2020).

so no material remains in the ground in an unsolidified state. Solidified chemicals, which are mostly neutral, are safer than the materials themselves and stable in the ground.

There have been no reports of water quality exceeding effluent standards from an estimated 200,000 construction sites employing this method conducted over approximately 30 years from the mid-1970s. It can be assessed, therefore, that the impact on groundwater due to chemical injection is not at a level that would cause environmental problems (Japan Grouting Association, 2015). Furthermore, chemicals do not contaminate the soil either, because the chemicals used in this method contain no heavy metals or harmful substances.

2.2.4 Long-Term Durability of Chemicals

In recent years, the chemical grouting method has come to be used for permanent purposes such as liquefaction countermeasures, seismic reinforcement of facilities, and improvement of the bearing capacity of the ground. Therefore, the long-term durability of the ground improved by this method is required. Japan Grouting Associations and Coastal Development Institute of Technology showed 5-year and 11-year in-situ long-term durability test results that indicated no deterioration of the improved ground. Based on these test results, the strength of the ground improved by water glass-based chemicals will either maintain its initial value or gradually increase over a period of five years (Japan Grouting Association, 2015). It has also been confirmed, furthermore, that solution type chemicals of activated silica provided 11-year strength of the improved ground larger than the initial strength (Coastal Development Institute of Technology, 2020).

2.3 PRELIMINARY SURVEY FOR CHEMICAL GROUTING

The most important aspect of the preliminary survey for the chemical grouting method is its applicability to the ground to be improved. In the survey, various investigations and tests are conducted to evaluate whether the chemical grouting method can make improved ground with designated characteristics. Investigations and tests listed in Table 2.1 are required as a preliminary survey not only for applicability of this method, but also for coping with the surrounding environment and existing underground structures.

The improvement effect of the chemical grouting closely depends on the grain size distribution of in-situ soil. For example, if the fine content of the ground is high, chemicals mainly penetrate in leaf vein shape, and uniform improvement cannot be expected. Again, when injection into sandy ground with a large uniformity coefficient, the chemicals will be affected by dilution, making it impossible to obtain the expected strength of the improved body. To obtain the designated properties of improved soil (strength and impermeability), chemicals should be injected by uniform permeation at a given concentration rather than penetration in leaf vein shape. Thus, it is important to evaluate the soil characteristics of the ground in design, based on proper preliminary investigation.

2.4 DESIGN

Figure 2.2 shows the design procedure of ground improved by the chemical grouting method. This section introduces settings of improvement specifications, injection specifications, and mix proportion design.

2.4.1 Setting of Improvement Specifications

Improvement specifications of the chemical grouting method consist of the strength of improved ground, improvement range, and improvement ratio. These specifications shall be set in accordance with the design

TABLE 2.1 Preliminary Investigation for Chemical Grouting

Target	Purpose	Contents	Importance[a]	Note
Soil layer, soil	Soil profile, category of soil	Sounding, sampling	†††	Soil boring log (geological columnar section), soil classification
	Dry density	Sounding	†††	*N*-value, etc.
	Grain size distribution (sand and fine content)	Grain size analysis	†††	Selection of injection method based on mechanism of grouting
	Compressive strength	Unconfined compression test or triaxial compression test	†	Cohesion
	Consistency	Test of physical properties	†	Fundamental information for estimation of improvement effect
	Size and content of gravel	Sampling for gravelly soil	††	Evaluation of drilling method and its efficiency
	Hydraulic property	In-situ test for hydraulic conductivity	†††	Hydraulic conductivity
Groundwater	Groundwater level	Measurement of groundwater level in borehole or well, measurement of pore water pressure using electric transducer in borehole	†††	Groundwater level or artesian head
	Water quality	Water quality analysis	††	Factor influencing on solidification and strength of chemicals, initial values for water quality management
	Groundwater flow	Flow layer logging, groundwater flow investigation	††	Groundwater flow layerFlow direction and velocity

(*Continued*)

TABLE 2.1 (CONTINUED) Preliminary Investigation for Chemical Grouting

Target	Purpose	Contents	Importance[a]	Note
Buried object	Buried object	Investigation of types of buried pipes	†††	Measures against damage to buried pipes
	Neighboring existing facility	Investigation of types and underground structure of facilities	†††	Measures against damage of displacement and deformation
Well and public waters[b]	Well	Investigation of location, depth, structure, purpose of use, usage situation	†††	Compliance with provisional guidelines
	Public water	Investigation of location, depth, shape, structure	†††	
Surrounding situation	Plants	Investigation of crops, plants, street, trees	††	Types of plants, season of sowing, plant, or harvest
	Life	Investigation of use of Buildings in Building Standards Act, neighboring existing schools or hospitals	††	Plan of time zone of construction work
	Traffic	Investigation of traffic volume, road width, possibility of detour	††	Plan of time zone of construction work and area of barrier for construction site

Source: Japan Grouting Association (2015).

[a] †††: Necessary, ††: Desirable, †: Optional.

[b] Public waters: Areas of water for public use such as rivers, lakes, ports and harbors, coastal seas, etc., including public waterways connected thereto, irrigation waterways, and other public-use waterways, excluding the public sewers and the river-basin sewers with a sewerage treatment plant (Ministry of Justice, Japan, 2023).

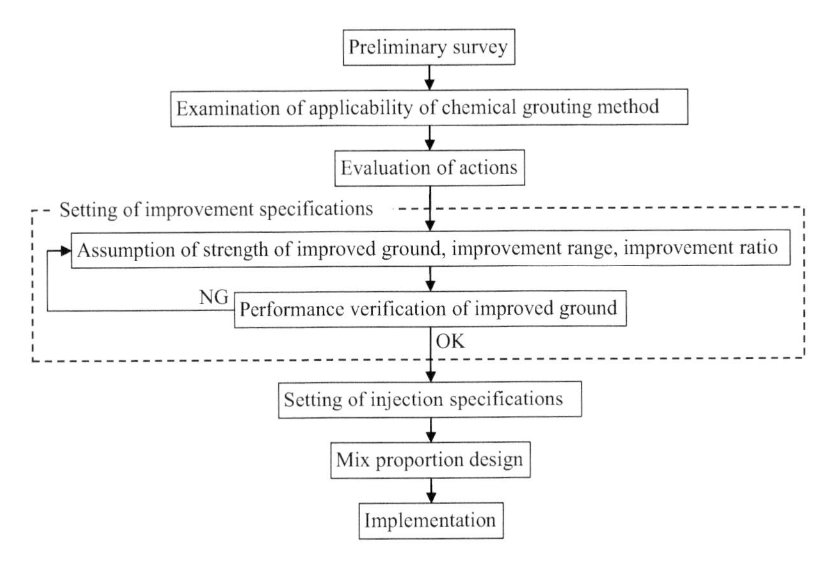

FIGURE 2.2 Procedure of chemical grouting method. (After The Overseas Coastal Area Development Institute of Japan, 2020).

concept and procedure exhibited in design guidelines and standards for each type of structure and facility. Many design guidelines and standards have recently adopted the reliability-based design method, in which specifications are determined to satisfy the performance requirements set in advance. Performance requirements are represented as an allowable degree of damage of the facility, such as, for example, its displacement due to earthquake motion. Improvement specifications as foundation ground of the facility are set to ensure that the facility meets performance requirements. Performance of the facility based on specifications determined is verified by model tests or numerical analysis as needed.

Improvement by the chemical grouting method brings about an increase of cohesion only and no increase of internal friction, as mentioned above. In case of sandy ground improved by uniform permeation rather than penetration in leaf vein shape, the increase of cohesion after improvement has been estimated to be 40 to 100 kN/m^2 depending on the N value of the original ground (Japan Grouting Association, 2015). On the other hand, when the purpose of the improvement is countermeasure against liquefaction, the performance of improved ground is verified by safety factor against liquefaction, which is defined as the ratio of liquefaction resistance of the ground to shear stress generated by earthquake motion (Coastal Development Institute of Technology, 2020).

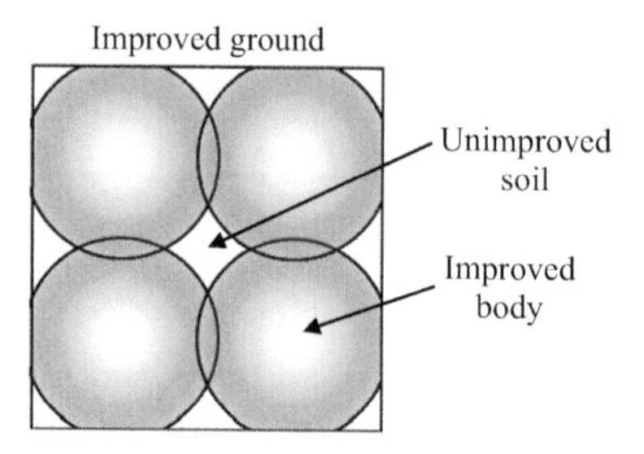

FIGURE 2.3 Image of improved ground with improved ratio below 100% (70%).

The improvement range is set as the size of improved ground necessary to achieve the purpose of improvement, such as increased imperviousness or strength of the ground. If there are multiple improvement purposes, the largest improvement range is employed in those required for each objective.

The improvement ratio is defined as a ratio of the net improved volume to the entire volume of the ground to be improved, that is, volume of the improvement range expressed as a percentage. The improvement ratio below 100% means that there remains some unimproved soil among improved bodies (Figure 2.3). The improvement ratio below 100% can be allowed when results of experiment or numerical analysis make certain that the performance requirement for the structure is satisfied, though the improvement rate shall be set at 100% in principle (Coastal Development Institute of Technology, 2020).

2.4.2 SETTING OF INJECTION SPECIFICATIONS

Injection specifications of the chemical grouting method consist of chemicals (type and its concentration), grouting ratio, grouting speed, grouting pressure, grouting interval, and gelling time. The injection specifications are set to achieve the above improvement specifications under the soil profile and construction conditions of the ground to be improved. The fine content has the greatest influence on injection specifications. The smaller the fine content of the ground, the easier the chemicals will be permeated, the grouting ratio, grouting speed, grouting interval, and grouting volume (improved diameter) will be relatively large, and the grouting pressure will be small.

Chemicals and their concentrations are selected, taking into consideration permeability, applicability to soil and groundwater, durability, and economic efficiency, in order to obtain the effect according to the purpose

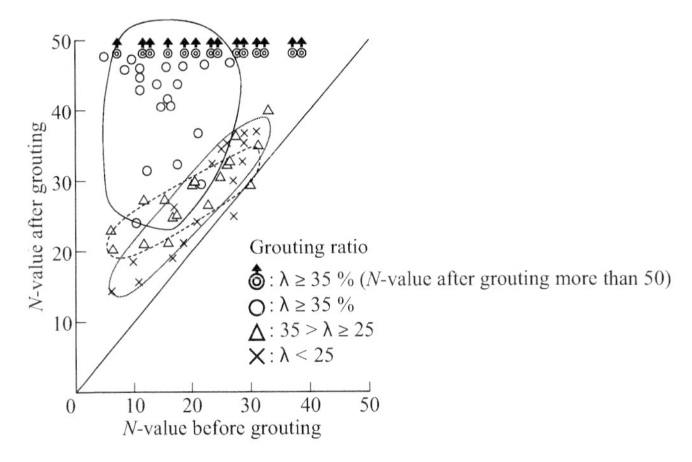

FIGURE 2.4 Relationship between grouting ratio and improvement effect. (After Japan Grouting Association, 2015).

of improvement and are decided based on the results of the mix proportion test described below.

The grouting ratio is defined as a ratio of the injected volume of chemicals to the volume of ground to be improved, expressed as a percentage. The injection rate greatly affects the improvement effect and is also an important factor for economic efficiency. Figure 2.4 shows the relationship between grouting ratio and improvement effect. This figure also indicates that the improvement effect on the strength of the ground (*N*-value) becomes significant when the grouting ratio exceeds 35%. The amount of chemicals to be injected is calculated with the following equation using the grouting ratio (Japan Grouting Association, 2015),

$$Q = \frac{V \lambda J}{100},$$

where Q is the amount of chemicals to be injected (m^3), V is the volume of ground to be improved (m^3), λ is the grouting ratio (%), and J is the importance factor. The importance factor is a kind of safety factor regarding the amount of injection and is set at 90 to 120% based on the importance of the construction work and required effect, but in normal cases it can be set at 100% (Japan Grouting Association, 2015).

High grouting speed improves the economic efficiency of work. On the other hand, high grouting speed tends to induce penetration in leaf vein shape, making it difficult to achieve uniform improvement. In design, the

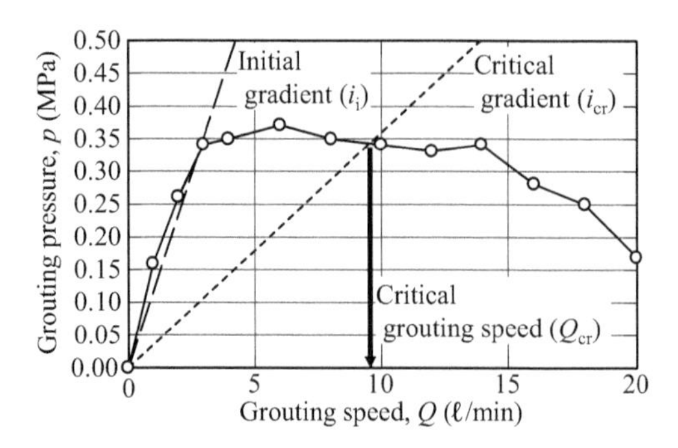

FIGURE 2.5 Example of result of critical grouting speed test. (After Coastal Development Institute of Technology, 2020).

grouting speed is often determined based on work experiences or in-situ test on the critical grouting speed. In the critical grouting speed test, the relationship between grouting speed and grouting pressure is obtained, as shown in Figure 2.5. The grouting pressure increases in conjunction with the grouting speed, and after it exceeds a certain grouting speed, the grouting pressure begins to decrease. This implies that grouting has changed from uniform permeation to penetration in leaf vein shape. To keep uniform permeation, a critical grouting speed indicated in Figure 2.5 is defined, which is used as the upper limit of grouting speed in construction management. In a case of small initial gradient of the grouting speed/grouting pressure relationship, there is also a method in which the grouting speed corresponding to the peak value of grouting pressure is set as the critical grouting speed.

The most important aspect of grouting pressure is its upper limit, as high grouting pressure can cause uplift of the ground surface and deformation of underground objects and nearby structures. In construction management, injection is conducted at a constant grouting speed, but injection speed is reduced if the pressure exceeds the upper limit. The value of the upper limit of grouting pressure depends on the injection method, so it should be set with reference to the technical manual for each method.

The grouting interval is closely related to placement of improved bodies, which must be arranged appropriately, based on the grouting ratio, grouting speed, and amount of chemicals determined from the injection operation time and the improvement range. A small grouting interval reduces the amount of chemicals per hole and improves reliability but increases the number of holes to be drilled and the cost. However, if the grouting interval is increased, the injection operation time becomes longer, making it difficult to manage the gelling time. The grouting interval may therefore have an upper limit, of which the value depends on the injection method. The upper limit can be set with reference to the technical manual for each method.

The gelling time is defined as the time from when solutions are mixed to when the chemicals lose fluidity. A gelling time that is too long results in dilution of chemicals by pore water. Then, improved bodies tend to have size larger and strength lower than expected. But, if the gel time is too short, penetrating in leaf vein shape is likely to occur.

2.4.3 Mix Proportion Design

The strength of soil improved by the chemical grouting method and the gelling time of chemicals are affected by the soil properties of the ground to be improved. The permeation grouting method requires laboratory tests for mix proportion design based on preliminary survey results (Coastal Development Institute of Technology, 2020). The laboratory tests consist of the strength test and the gelling time test.

The strength test is conducted for determination of the concentration of chemicals to obtain improved soil of designated strength in situ, and the strength is usually evaluated through unconfined compression tests. It is important to prepare specimens with a dry density equivalent to that of the target ground, referring to the results of in-situ tests, since the strength of improved soil is affected by the density of the ground.

The gelling time test is conducted to design mix proportion of chemicals, such as pH of chemicals and the amount of hardening agent to be added, in order to obtain the gelling time of chemicals equivalent to injection time on-site. Test sample must be in-situ soil, because the gelling time of chemicals is affected by the pH of the ground, the calcium contained in shell fragments, and the fine content.

2.5 IMPLEMENTATION

Construction should be planned, implemented, and managed based on the technical manual for each construction method, as the procedures and control items depend on the construction method (Railway Technical Research Institute, 2011; Japan Grouting Association, 2015; Coastal Development Institute of Technology, 2020).

Construction management mainly controls quantity, quality, and environmental conservation. Quantity control involves managing the chemicals and hardening agents used. Quality control verifies the validity of the design and manages the accuracy and effectiveness of construction. Environmental conservation involves monitoring the contamination of groundwater and public water, the impact on the existing nearby structures, and flora and fauna. This section explains quality control.

In quality control of the chemical grouting method, it is important to verify the design, which consists of improvement specifications, injection specifications, and mix proportion design, as explained in the previous section, through the test construction in situ. If necessary, the design should be modified to a suitable one for the ground to be improved, based on the result. Modified design suitable for the site must be accurately implemented on the site. For accurate construction, both drilling position and injection position must be monitored. Furthermore, for uniform permeation, it is also important to control the quantity, quality, grouting speed, and grouting pressure of the chemicals. To ensure the quality of improved ground, it should be evaluated by the post survey. If the purpose of improvement is to increase imperviousness of the ground, the permeability coefficient is evaluated through an in-situ permeability test. Improved ground with the purpose of increasing strength is evaluated on the strength or rigidity through laboratory tests or sounding. Unconfined compression tests or triaxial tests are performed on specimens sampled from improved ground as laboratory tests. For sounding, standard penetration tests, dynamic cone penetration tests, and pressuremeter tests are performed.

REFERENCES

Coastal Development Institute of Technology, 2020. *Technical Manual for Permeation Grouting Method* (2nd Edition). 10–26, 42–56, 83–92, 123–125, 136–137. (in Japanese).

Japan Grouting Association, 2015. *Chemical Grouting Method* (Revised Edition). 3–4, 8–22, 222–263, 280–306, 338–340. (in Japanese).

Matsuoka, Masatada, 2021. "Experiments on Water Glass," *Chemistry and Education*. 69. 2. 58–59. (in Japanese).

Ministry of Justice, Japan, 2023. "Water Pollution Prevention Act, Act No.138 of 1970: Japanese Law Translation." https://www.japaneselawtranslation.go.jp/ja/laws/view/2815

Railway Technical Research Institute, 2011. *Manual for Design and Implementation of Chemical Grouting*. 13–38, 46–53, 66–73, 169–174. (in Japanese).

The Overseas Coastal Area Development Institute of Japan, 2020. *Technical Standards and Commentaries for Port and Harbour Facilities in Japan*. 544–545, 829–834. (in Japanese).

CHAPTER 3

Performance Verification of Liquefied Ground at Airports

Teruhisa Fujii

3.1 INTRODUCTION

In recent years, disasters such as earthquakes and tsunamis have occurred frequently in Japan and other parts of the world, and there is a strong need for countermeasures against such disasters, including earthquake countermeasures.

In August 2005, the Civil Aviation Bureau, Ministry of Land, Infrastructure, Transport and Tourism (MLIT) established the "Committee for Earthquake Resistant Airports" with a panel of experts to review the ideal airports in the event of an earthquake and to discuss the basic concept of improving earthquake resistance.

In response to this, the *Guidelines for Airport Seismic Design* (Civil Aviation Bureau, Ministry of Land, Infrastructure, Transport and Tourism, 2022) presents the concept of seismic performance of each facility according to the functions required of airports, as well as the basic approach for verifying the seismic performance of each facility.

However, in the ground with variations in ground constants, liquefaction may occur in areas that do not satisfy the required liquefaction strength for the earthquake motion to be designed when an earthquake

DOI: 10.1201/9781032670133-3

of the design earthquake motion level occurs, and it is necessary to verify whether the basic airport facilities satisfy the design limits in such cases.

However, it is difficult to evaluate the performance of basic airport facilities on such soils using conventional specifications, and it is necessary to verify the performance of the improved soils using new methods based on performance specifications. This chapter summarizes the seismic performance of airport facilities according to the functions required of airports and then presents the flow of performance verification based on the heterogeneity of the ground.

3.2 EXAMPLES OF DAMAGE CAUSED BY LIQUEFACTION

3.2.1 Overview of Past Liquefaction Damage

So far, there are 22 airports in Japan that were damaged by the earthquake. The damage occurred in paved areas, buildings, seawalls, drainage facilities, and community ditches. Among them, liquefaction was confirmed at four airports: Niigata Airport, Yonago Airport (Miho Airfield), Matsuyama Airport, and Sendai Airport, as shown in Table 3.1.

This section provides a detailed overview of the liquefaction damage at Niigata Airport.

In October 2007 a full-scale experiment was conducted at Ishikari Bay New Port in Hokkaido, Japan, to reproduce a full-scale airport facility and to simulate liquefaction of the ground by controlled blasting to understand the behavior of the ground during liquefaction. An overview of the experiment is also presented.

TABLE 3.1 Airports Damaged by Earthquakes

Airport	Earthquake	Damages
Niigata	Niigata earthquake (1964)	Pavement, Seawalls, Drainage facilities, Buildings
Yonago	Tottori Western earthquakes (2000)	Pavement, Revetment
Matsuyama	Geiyo earthquake (2001)	Landing Zone
Sendai	Tohoku Area Pacific Ocean earthquake (2011)	Pavement, Drainage Facilities, Fire and Water Facilities, Railroad Bridges, Railroad Walls

FIGURE 3.1 Damage at Niigata airport.

3.2.2 Damage at Niigata Airport

The following is an overview of the damage caused by liquefaction at the basic airport facilities at Niigata Airport during the Niigata Earthquake (1964) (Figure 3.1) (Sugano et al., 2012).

Most of the damage to Niigata Airport caused by the Niigata Earthquake was due to liquefaction of the sandy soil. The airport's basic facilities, such as runways, taxiways, and aprons, were severely damaged by destruction, cracks, subsidence, and groundwater eruptions, and the airport was completely paralyzed. Fault cracks were observed in the area of about 300 m from the southern end of Runway A.

This occurred at the boundary between the former dune area and the floodplain. There was little damage south of that boundary and significant damage to the north. The pavement slab was found to have failed due to uplift of the pavement near the point of change in longitudinal slope at approximately 500 m from the southern end. Wavy unevenness was observed on the pavement surface due to unequal settlement at approximately 800 m on the north side.

A staircase fault occurred about 150 m from both ends of runway B. This was assumed to be the boundary between the coastal dune and the flood plain, but the pavement on the coastal dune side was completely cracked and destroyed. The runway pavement surface was uneven due to unequal settlement.

Sand (with mud or silt) eruptions occurred, along with groundwater eruptions throughout the landing zone, except in the former dune area.

3.2.3 Overview of On-Site Full-Scale Experiments at Ishikari Bay New Port, Hokkaido

In order to investigate reasonable liquefaction countermeasures for runways constructed on soft ground in reclaimed coastal areas, a full-scale runway specimen was constructed at Ishikari Bay New Port in Hokkaido, Japan, and field experiments were conducted in which controlled blasting was used to actually cause liquefaction of the ground.

The runway, which is the main facility of the airport, was constructed in a 50 m × 60 m area within the 1.65 ha experimental yard, with a pavement structure that can accommodate the takeoff and landing of a Boeing 747-class aircraft. In the runway pavement area, three methods were used for liquefaction countermeasures: Compaction grouting method (CPG), Permeation grouting method, and Ultra multi grouting method (Figure 3.2).

FIGURE 3.2 Major airport facilities constructed on-site.

In order to explore the possibility of cost reduction, standard designs for each method and measures with different improvement rates and depths were implemented.

The stratigraphic composition of the subject ground is, in order from the surface, a buried soil layer with an N value of 1 to 8, a sandy soil layer with an N value of 3 to 12, and a sandy soil layer with an N value of 8 to 20 below the buried soil layer. The groundwater level was found to be generally in the range of GL-2.0 m to 2.5 m from the water level in the borehole. The assumed liquefaction layer in the experimental field is up to GL-10 m, and the explosive locations were designed based on the results of preliminary experiments to be loaded in a 6.5 m pitch grid (approximately 4 kg per location) at two depths: GL-4.5 m and -9.0 m (Figure 3.3).

To simulate liquefaction directly under the runway, a bent borehole was used after the pavement work was completed, and the pavement was loaded with chemicals. After drilling, loading operations were repeated in each hole, and a total of 583 emulsion-type water-containing explosives

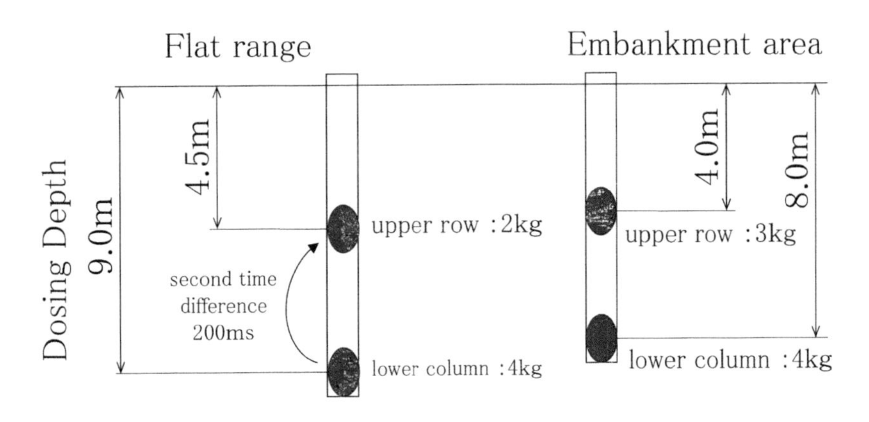

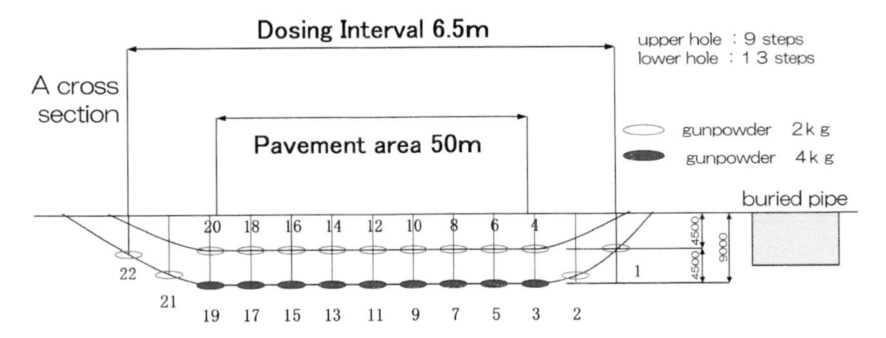

FIGURE 3.3 Charging cross section for controlled blasting.

with a total explosive weight of 1760 kg were buried. During the experiment, blasting was performed at approximately 0.2-second intervals, and liquefaction of the ground was reproduced by staged blasting over a total blasting time of 139 seconds. In the experiment, pore pressure measurements were used to measure liquefaction conditions and the dissipation process of excess pore pressure during blasting, and pre- and post-experimental leveling surveys confirmed ground subsidence conditions and deformations such as unequal settlement of the facility. The reproduction of liquefaction conditions is aimed at the increase in excess pore water pressure caused by the collapse of the sand particle skeleton around the hypocenter due to the vibration caused by the explosion, which is different from the mechanism of the increase in excess pore water pressure caused by negative dilatancy due to shear deformation of sand and soil during earthquakes in general. In particular, the effects of seismic vibration on the facility cannot be reproduced by liquefaction experiments using controlled blasting. Shaking table experiments, laboratory soil tests, and numerical simulations were also conducted.

Blasting was conducted on October 27, 2007 (11:00 ignition). No cracks were observed in the runway pavement immediately after blasting, and sand and water fountains due to liquefaction were observed at various locations in the surrounding area (Figure 3.4).

FIGURE 3.4 Sand eruption immediately after the experiment.

The records from the pore pressure gauges installed underground also confirmed that the assumed liquefied layer had reached a state of full liquefaction. The settlement of the liquefied area directly under the runway on the seventh day after blasting was found to be extremely small, while settlement of more than 30 cm was observed in the unconsolidated area. Because the sand in the liquefied layer at the site is relatively uniform fine sand, the settlement of the liquefied layer at the site was about 80% within four hours after blasting, and it was almost settled in about seven days. Since the sand quality of the assumed liquefaction layer at each airport is different, the time required for dissipation of excess pore water pressure is estimated to range from a few hours to several days, depending on the difference in permeability.

The contents of this experiment are summarized below:

1. Various airport facilities were deformed due to dissipation of excess pore water pressure after liquefaction. Airport facilities such as asphalt pavement, approach lights, localizers, and glide slopes, all of which were constructed in unimproved areas, suffered from settlement or unequal settlement. In addition, severe sand eruptions around manholes on the sides of asphalt pavements were observed, which seriously affected the functionality of some facilities.

2. The cost reduction by reducing the improvement ratio and the improvement area, and the effect of the ground improvement at that time were examined. As a result, it was confirmed that the liquefaction countermeasure method based on compaction can reduce the area of extra improvement and the improvement ratio—and that partial improvement in the depth direction is possible in the chemical injection method.

The extra improvement is to mitigate the intrusion of excess pore water pressure from the unprotected area into the liquefied area, and was established based on the study on liquefaction countermeasures behind the quay wall in the port. In the case of partial improvement, uniform settlement is acceptable from the viewpoint of the runway function, and the entire liquefaction layer should be improved in principle, but depending on the site conditions, it may be possible to leave an unprotected area in the lower part. In both cases, it is necessary to analyze the behavior during a hypothetical earthquake using numerical simulation and confirm that

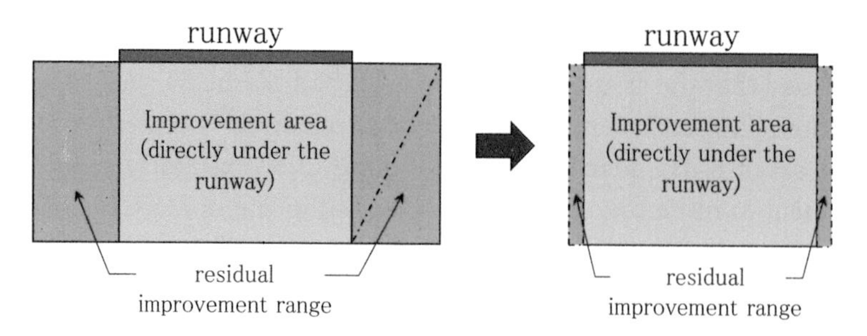

FIGURE 3.5 Image of reduced remaining improvement width.

the required performance (flatness, specified slope, and bearing capacity) is ensured (Figure 3.5).

3.3 PERFORMANCE REQUIREMENTS AND PERFORMANCE SPECIFICATIONS FOR AIRPORTS

3.3.1 Roles Required of Airports According to the "Committee for the Study of Airports Resistant to Earthquakes"

Airports that can serve as emergency transportation hubs are required to function as a base for emergency and lifesaving activities immediately following a disaster and to be capable of receiving emergency supplies and personnel within three days after a disaster. In addition to the above, "airports of air transport importance" are required to be capable of operating regular commercial flights within three days after the disaster, to have a transportation capacity equivalent to 50% of normal operations as soon as possible after resumption of operations in order to halve the economic damage caused by the earthquake, to maintain the air network, to ensure the sustainability of economic activities in the surrounding area, and to maintain the functions of the capital city. The resumption of flights is expected to halve the economic damage caused by the earthquake. This section introduces the contents of the study conducted by the "Committee for Earthquake Resilient Airports."

1. Role of Airports Required in the Event of Earthquake Disasters and Basic Approach to Improving Earthquake Resistance

   Airports are required to play a role as a base for transporting emergency supplies and personnel in the event of an earthquake disaster, as evidenced by the role of airports as a base for transporting emergency supplies and personnel in the event of a major earthquake such as the Niigata Chuetsu Earthquake. In particular, when the airport plays an important role in the network, and when there is a risk that a functional failure of the airport could affect aircraft operations nationwide, or when the airport is located in an area with active economic activity in the surrounding area and the economic loss due to a functional failure would be significant, the airport is required to maintain the aviation network and ensure the continuity of economic activity in the surrounding area. In order to fulfill these roles, airport facilities should be designed and constructed in a manner appropriate for the area and have the following basic earthquake resistance requirements.

- The earthquake ground motion will not significantly interfere with the functions necessary for aircraft operation.
- No significant impact on human life in the event of large-scale seismic motion.
- Air traffic control functions do not stop for the safe operation of aircraft in the event of large earthquakes.

In addition, the airport should have the functions shown in Table 3.2, depending on the role required of the airport in the event of an earthquake disaster.

2. Measures to Improve Earthquake Resistance of Airport Facilities, etc.
   At airports that serve as hubs for emergency transportation, measures are being considered to improve the seismic resistance necessary for the transport of emergency supplies and personnel by Self-Defense Forces aircraft and others. In addition, at airports that are important for air transport, measures to improve earthquake resistance are considered necessary to ensure 50% of the normal volume of scheduled civilian air transport. Table 3.3 shows the measures to be taken to improve the seismic performance of basic airport facilities.

3. Airport Operations in the Event of an Earthquake Disaster
   In the event of an earthquake, airports must not only be earthquake-resistant, but must also have sufficient software measures in place to ensure that they can fulfill their roles as transportation hubs for emergency supplies and personnel, and to maintain the airline network and economic activities in the surrounding area.

TABLE 3.2 Functions Required of Airports

Role of Airports	Functions Required of Airports	Basic Approach to Maintenance to Ensure Functionality
Airports that serve as hubs for emergency transportation	• Base function for emergency and lifesaving activities (in the very early stage after a disaster). • Acceptance function for emergency supplies and personnel transport (within three days after a disaster).	• Conduct a detailed seismic survey and accurate damage prediction for airport facilities and take necessary countermeasures based on the survey. • Airports with a runway of about 2000 m and capable of receiving mass transport by Self-Defense Forces aircraft should have their seismic capacity improved for this purpose. For other airports, it is necessary to improve the seismic resistance of facilities for transportation by helicopters and small aircraft. • In the event of an earthquake disaster, necessary countermeasures should be taken in advance if there is a risk of secondary damage.
Important airports for air transportation	• Capability to operate regular commercial aircraft for up to three days after the disaster. • Aiming to halve the economic damage caused by the earthquake disaster, we will secure transportation capacity equivalent to 50% of normal operations at the earliest possible stage after the resumption of operations. • Maintaining the airline network and ensuring the continuity of economic activities in the hinterland and maintaining the functions of the capital.	• To improve the seismic resistance of runways and taxiways, etc., as necessary to ensure the transport capacity equivalent to 50% of the normal capacity of scheduled commercial aircraft at an early stage as much as possible.

TABLE 3.3 Measures to Improve Earthquake Resistance to be Taken at Each Facility

Target Facilities	Measures to Improve Earthquake Resistance
Runway	• Airports that serve as hubs for emergency transportation must be earthquake-resistant with a runway length (approximately 2000 m) that allows for the takeoff and landing of Self-Defense Force transport aircraft (C-1 transport aircraft, C-130 transport aircraft, etc.). • At airports vital to air transport, in addition to being able to take off and land Self-Defense Force transport aircraft, it is necessary to have the same operational configuration as under normal conditions and to ensure that the number of runways and their total length necessary to secure 50% of transport volume are earthquake-resistant so they can be operated using air security facilities. • For landing zones, etc., it is necessary to prepare in advance a restoration system that will allow for the clearing of land within three days after the disaster.
Taxiway	• Airports that serve as hubs for emergency transportation must ensure the earthquake resistance of taxiways that allow Self-Defense Force transport aircraft and other aircraft to travel between the runway and the parking area. • At airports important for air transportation, it is necessary to reduce the time that aircraft occupy the runway and secure throughput, in addition to allowing Self-Defense Force aircraft to travel between the runway and the parking area. It is also necessary to secure the earthquake resistance of the parallel taxiways and the terminal attachment taxiways that connect to them.
Aprons, etc.	• Airports that serve as hubs for emergency transportation should ensure the seismic resistance of their parking areas based on the scale of emergency transportation activities, etc., so that Self-Defense Force transport aircraft and other aircraft can be parked there. • At airports important for air transport, in addition to being able to park Self-Defense Force transport aircraft and other aircraft, it is necessary to avoid delays and confusion due to lack of parking space and to ensure that the parking space is earthquake-resistant enough to accommodate the traffic volume.

3.3.2 Functions Required of Airports According to *Guidelines for Airport Seismic Design* (Civil Aviation Bureau, Ministry of Land, Infrastructure, Transport and Tourism, 2022)

For an airport to function properly, the seismic performance of each facility must be determined according to the required functions of the airport (transportation functions that should be secured even in the event of an earthquake disaster). The seismic design guidelines for airports provide the basic concept of the seismic performance of each facility according to the required functions of the airport and the basic concept of the verification of the seismic performance of each facility.

1. Basics of Seismic Design
   In designing airport civil engineering facilities, it is necessary to ensure that the seismic performance is appropriate for the required functions of the airport, and that the following basic seismic performance is provided for both level 1 and level 2 earthquake motion, regardless of the functions of the airport.

   - The earthquake ground motion shall not affect the functions necessary for aircraft operation.
   - The earthquake motion shall not seriously affect human lives, properties, or socioeconomic activities.

2. Seismic performance of airport civil engineering facilities according to the type of transportation
   In designing airport civil engineering facilities, it is indicated that each facility constituting the airport should have seismic capacity to cope with the expected transportation patterns after an earthquake.

   - Some airport civil engineering facilities may directly affect human lives due to damage to the facilities themselves, while others may prevent aircraft operations and affect human lives, property, or social activities due to damage to the facilities. For other facilities, seismic performance is required according to the transportation function that should be secured immediately after the earthquake (about three days after the earthquake).
   - Depending on the transportation function to be secured after an earthquake, various types of transportation are assumed, such as passenger transportation by fixed-wing aircraft, emergency cargo transportation, and emergency cargo transportation by rotary-wing aircraft. In general, the seismic performance of airport civil engineering facilities should be higher for passenger transport than for emergency cargo transport, and it should be noted how much seismic performance is required for each type of transport. Although emergency transport aircraft capable of

mass emergency transportation can generally take off and land on shorter runways than is possible for commercial aircraft, it is necessary to consider the geographical conditions and transportation modes.

3. Seismic performance of airport civil engineering facilities

In designing airport civil engineering facilities, it is indicated that the basic seismic performance required of airports after an earthquake and the seismic performance required according to the transportation function shall be set based on the earthquake scale and facilities, and that the evaluation items for these performances shall be set appropriately according to the earthquake scale and structural characteristics of the facilities.

- Airport facilities must be able to continue to operate without loss of functionality in the event of a level-1 earthquake motion. In the case of level-2 earthquake motion, it is necessary to ensure the repairability of facilities that may have a serious impact on human lives, properties, or socioeconomic activities due to damage.
- Other facilities must be able to continue to operate without loss of functionality in the event of a level-1 earthquake motion. In addition, in order to ensure the overall seismic performance of the airport as a whole, the structure of each facility must be strong enough to withstand level-2 earthquake motion if it is technically feasible and economically reasonable for the facility to continue to be used, i.e., if it is required to be repairable.
- The evaluation items for the seismic performance of airport civil engineering facilities are generally shown in Tables 3.4 and 3.5.

TABLE 3.4 Seismic Performance of Major Airport Civil Engineering Facilities Against Level-1 Earthquake Motion (Evaluation Items)

Facilities	Evaluation Item
Runway	1. Whether the ground is liquefied or not
Taxiway	2. Ground deformation (slope and step)
Apron	3. Bearing capacity of the ground
Over-run zone	4. Crack development status

TABLE 3.5 Seismic Performance of Major Airport Civil Engineering Facilities against Level-2 Earthquake Motion (Evaluation Items)

<table>
<tr><th rowspan="2">Facilities</th><th colspan="3">Forms of Transportation Required after a Major Earthquake</th></tr>
<tr><th>By Fixed-Wing Aircraft Passenger Transport</th><th>By Fixed-Wing Aircraft Transportation of Emergency Supplies</th><th>Transport of Emergency Supplies by Rotary-Wing Aircraft</th></tr>
<tr><td>Runway</td><td colspan="2" rowspan="3">1. Whether the ground is liquefied or not
2. Ground deformation (slope and step)
3. Bearing capacity of the ground
4. Crack development status</td><td>1. Ground deformation</td></tr>
<tr><td>Taxiway</td><td rowspan="2">1. Whether the ground is liquefied or not
2. Ground deformation (slope and step)
3. Bearing capacity of the ground</td></tr>
<tr><td>Apron</td></tr>
<tr><td>Over-run zone</td><td colspan="3">—</td></tr>
</table>

3.4 NECESSITY OF PERFORMANCE VERIFICATION AND VERIFICATION METHOD

3.4.1 Verification Method Shown in the Seismic Design Guideline

The seismic performance of airport civil engineering facilities should be verified based on the seismic response analysis method according to the structural characteristics of the facilities. For facilities that require consideration of liquefaction, liquefaction determination should be performed based on an appropriate method, and the liquefaction effects should be considered in the design if necessary.

The design limits for performance verification are to be set appropriately according to the required performance of the facility in question, the scale of the earthquake, and its structural characteristics. In the *Guidelines for Airport Seismic Design* (Civil Aviation Bureau, Ministry of Land, Infrastructure, Transport and Tourism, 2022), the performance limits required for airports at the time of a large-scale earthquake are to ensure the slope specified in Article 79, Paragraph 1, Item 3 of the Civil Aeronautics Law Enforcement Regulations and to ensure the necessary ground bearing capacity for the aircraft loads expected at the airport.

The maximum longitudinal slope of a runway is specified for each runway length and distance from the end of the runway, as shown in Table 3.6.

The maximum cross slope is specified for each wingspan of an aircraft using a land airport, as shown in Table 3.7.

TABLE 3.6 The Specified Value of the Maximum Longitudinal Slope of the Runway

Runway Length		Less than 800 m (%)	800 m to 1200 m (%)	1200 m to 1800 m (%)	Over 1800 m (%)
Maximum longitudinal gradient	The portion from the end of the runway to a distance of 1/4 or less of the runway length	2.0	2.0	1.5	0.8
	Other than the above	2.0	2.0	1.5	1.25

TABLE 3.7 Specified Maximum Cross-Gradient of the Runway

Wingspan of the Subject Aircraft	Maximum Cross-Gradient (%)
Less than 15 m	2.0
More than 15 m but less than 24 m	2.0
More than 24 m but less than 36 m	1.5
More than 36 m but less than 52 m	1.5
More than 52 m but less than 65 m	1.5
More than 65 m but less than 80 m	1.5

3.4.2 Verification Method Adopted in this Document

Figure 3.6 shows the flow of performance verification in the design of basic airport facilities.

Figure 3.6 shows the performance verification for a post-improvement ground with liquefaction countermeasures applied to a liquefaction-prone ground. The “local liquefaction” shown in Figure 3.6 is an event that occurs during earthquakes in areas of soil improved by the chemical injection method that do not meet the design base strength.

In the first part of this document, “a) Bearing Capacity Verification Bearing Capacity Analysis Considering Strength Heterogeneity (Including Local Liquefaction),” shear strength is expressed by random field theory for improved soil to prevent liquefaction, and bearing capacity analysis considering strength heterogeneity is performed by Monte Carlo simulation using the finite element method and shear strength reduction method.

Next, “b) Deformation verification: Deformation analysis with dissipation of excess pore water pressure considering strength heterogeneity (including local liquefaction)” is based on the analysis of deformation behavior associated with dissipation of excess pore water pressure after

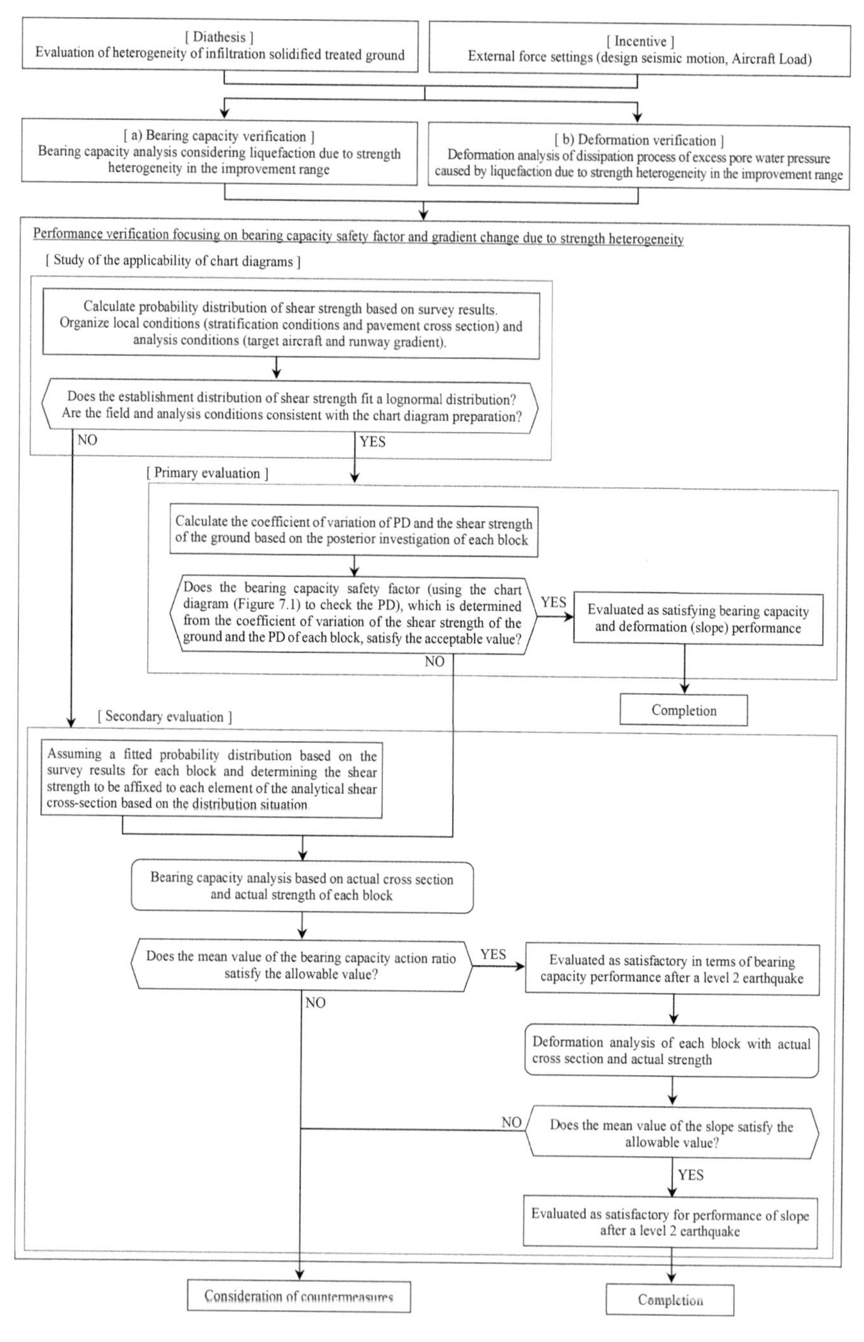

FIGURE 3.6 Flow of performance verification of improved ground by chemical grouting in this chapter.

liquefaction of ground improved to prevent liquefaction by expressing shear strength using random field theory and modeling simply whether the ground liquefies during earthquakes.

However, these methods require a large amount of computation time for a single condition, and it is not realistic to perform the analysis for all the soils to be improved in the future. Therefore, we attempted to evaluate the performance of bearing capacity and deformation by creating chart diagrams focusing on the compliances and coefficients of variation. The compliance ratio is the ratio of the strength that satisfies the design basis strength to the probability distribution of the shear strength, which is assumed to be a log-normal distribution, and is inversely related to the failure rate. Using this chart, it is possible to easily determine whether or not the bearing capacity and deformation performance are maintained by understanding the compliance rate and variation of the unconfined compression strength of the improved soil, without conducting a detailed bearing capacity analysis or deformation analysis.

REFERENCES

Civil Aviation Bureau, Ministry of Land, Infrastructure, Transport and Tourism, 2022. *Design Guidelines and Design Examples for Airport Civil Engineering Facilities (Seismic Design Version).*

Sugano, T., Nakazawa, H., Kohama, E., 2012. "Case studies on estimation of liquefaction damage by geophysical exploration in airport site," Technical Note of the Port and Airport Research Institute, No. 1247, 4.

CHAPTER 4

Liquefaction Strength Evaluation for Improved Ground by Chemical Grouting

Yasutaka Kimura

4.1 INTRODUCTION

Liquefaction of sandy soil is one of the causes of damage to civil engineering structures such as ports, airports, river embankments, and road embankments during recent large-scale earthquakes. Ground improvement methods for the purpose of liquefaction countermeasure of sandy soil can be broadly classified into densification by compaction, pore water pressure dissipation method, and solidification methods. Among these methods, the consolidation method using solidifiers or chemical solutions causes large variations in soil constants such as modulus of elasticity and shear strength constants in the improved ground due to heterogeneity in mixing conditions of solidifiers and chemical solutions, and heterogeneity of the target soil (The Ports and Harbours Association of Japan, 2018; Coastal Development Institute of Technology, 2020).

The current quality control method is based on taking undisturbed samples of ground improved by the chemical grouting (hereinafter referred to as "improved ground") and confirming the unconfined compression strength q_u by an unconfined compression test, as described in

DOI: 10.1201/9781032670133-4

the *Technical Manual for Permeation Grouting Method* (Coastal Development Institute of Technology, 2020).

This is based on the fact that in the seismic design system of the chemical grouting, there is a proportional relationship between the liquefaction strength ratio $R_{L20,5\%}$ (dynamic strength) and the unconfined compression strength q_u (static strength) of the improved ground, as shown in the black plot in Figure 4.1.

The black plots in Figure 4.1 are the results of cyclic triaxial and unconfined compression tests (Coastal Development Institute of Technology, 2020) using laboratory mix samples that were not affected by disturbance. Therefore, it is necessary to obtain a highly accurate unconfined compression strength q_u with less influence of disturbance of the specimen in order to confirm the strength of the improved ground.

However, since the unconfined compression strength q_u of the improved ground is very low, ranging from 50 to 100 kN/m², there is a possibility of sample disturbance during sampling, sample transportation, and sample molding—and a decrease in effective confining pressure due to stress release.

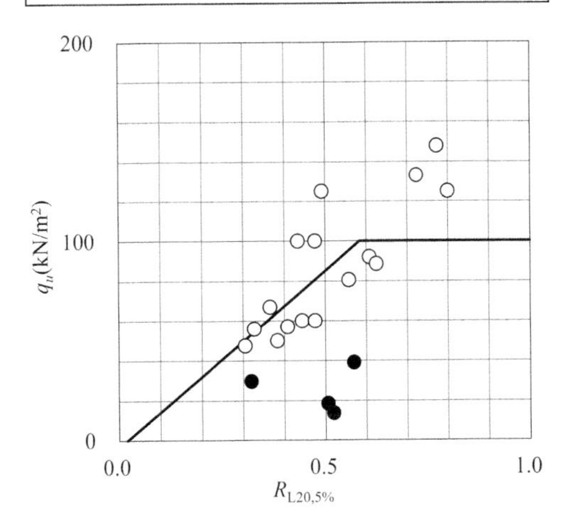

FIGURE 4.1 Relationships between q_u and liquefaction strength ratio.

The black plot in Figure 4.1 shows the relationship between the unconfined compression strength q_u and liquefaction strength ratio of samples collected from the improved ground tested at Site A using a triple-tube sampler, which is generally known as a sampling method with less disturbance.

This may be due to the fact that the measured unconfined compression strength q_u is lower in unconfined compression tests, which are easily affected by turbulence without confining pressure, than in triaxial tests where confining pressure is applied, because of turbulence during sampling, even if samples are collected using a triple-tube sampler.

For this reason, the *Technical Manual for Permeation Grouting Method* (Coastal Development Institute of Technology, 2020) states that improvement effectiveness should be evaluated by silica content tests and in-situ tests, which are not affected by disturbance.

However, Kawamura et al. (2002) confirmed the correlation between the unconfined compression strength q_u in improved ground that had been implemented and the yield pressure from a borehole load test and samples taken by rotary triple-tube sampling, and considered that the unconfined compression strength q_u may have been underestimated because it includes the effect of sample disturbance.

Okada et al. (2012) reported a case study on the investigation and evaluation of the unconfined compression strength q_u of improved ground by the rapid infiltration method (super multi-point injection method) (Okada and Takematsu, 2012), which is one of the chemical groutings, utilizing borehole load test and in-situ tests using PS logging, indicating the necessity of evaluation by in-situ tests, etc., instead of unconfined compression tests.

This background necessitates the establishment of an alternative investigation and strength verification method to evaluate the unconfined compression strength q_u, which is unaffected by specimen disturbance.

Sugano et al. (2020) proposed the in-situ borehole load test and the piezoelectric drive cone test (PDC test) as an alternative method for evaluating the unconfined compression strength q_u, of improved ground after verifying their applicability through laboratory model tests and field investigations.

However, there have been no cases in which the geotechnical investigation methods were combined based on the characteristics of the ground and each investigation method to evaluate the workmanship and quality of the improved ground at the same site.

In this chapter, a strength evaluation method that combines in-situ test and laboratory tests is developed as a method less susceptible to disturbance of specimens due to stress release and disturbance during sampling, and examples of improved ground evaluation using the proposed strength evaluation method are presented.

However, the correlation between the unconfined compression strength q_u obtained from conventional unconfined compression tests and the unconfined compression strength q_u estimated from other methods of investigation and strength verification is also presented, along with the results of the verification after comparing each method.

4.2 STRENGTH VERIFICATION METHOD AND APPLICATION CONDITIONS OF IMPROVED GROUND

In the evaluation of the improvement effect of improved ground, it is confirmed that the liquefaction strength ratio required for the actual ground is satisfied. However, repeated triaxial tests to determine the liquefaction strength ratio require four or more specimens (Japanese Geotechnical Society, 2020), so it is necessary to take a series of undisturbed samples of a 1 m section of the improved ground under the same conditions. However, the unconfined compression test can determine the unconfined compression strength q_u with a sample of about 0.2 m.

In this section, in-situ test and laboratory tests are used to estimate the strength of the improved ground.

4.2.1 Strength Evaluation by In-Situ Tests

4.2.1.1 Estimation of Unconfined Compression Strength q_u by Piezoelectric Drive Cone Test (PDC Test)

The PDC test is a sounding test that uses a tip cone to estimate the fine fraction content F_c and liquefaction strength from the penetration resistance N_d, which corresponds to the N-value, and the excess pore water pressure in the ground generated by the impact penetration (Yoshizawa et al., 2013). The shear strength of a sandy soil injected with chemicals is considered to be the shear strength due to the internal friction angle before chemical grouting plus the cohesion due to the solidification of the chemicals in the pores of the sandy soil (Suwa et al., 2006).

If the shear resistance with added cohesion due to solidification can be measured and confirmed as an increment of N-value, confirmation of the unconfined compression strength q_u and the increment of N-value ΔN

(N-value after improvement – N-value before improvement) can be obtained. As a result, the in-situ unconfined compression strength q_u of the soil can be estimated from the ΔN values before and after chemical grouting.

Figure 4.2(a) shows the relationship between the ΔN_d value from the PDC test and the unconfined compression strength q_u from the unconfined compression test (Ports and Harbours Bureau, Ministry of Land, Infrastructure, Transport and Tourism, 2018) conducted on the improved ground at the actual site (Site B) in the past, and the case of Site A is shown in addition. The unconfined compression strength q_u in the case of Site A is based on an unconsolidated undrained triaxial test on a less disturbed sample taken against the improved ground. As shown in Figure 4.2(a), there is a

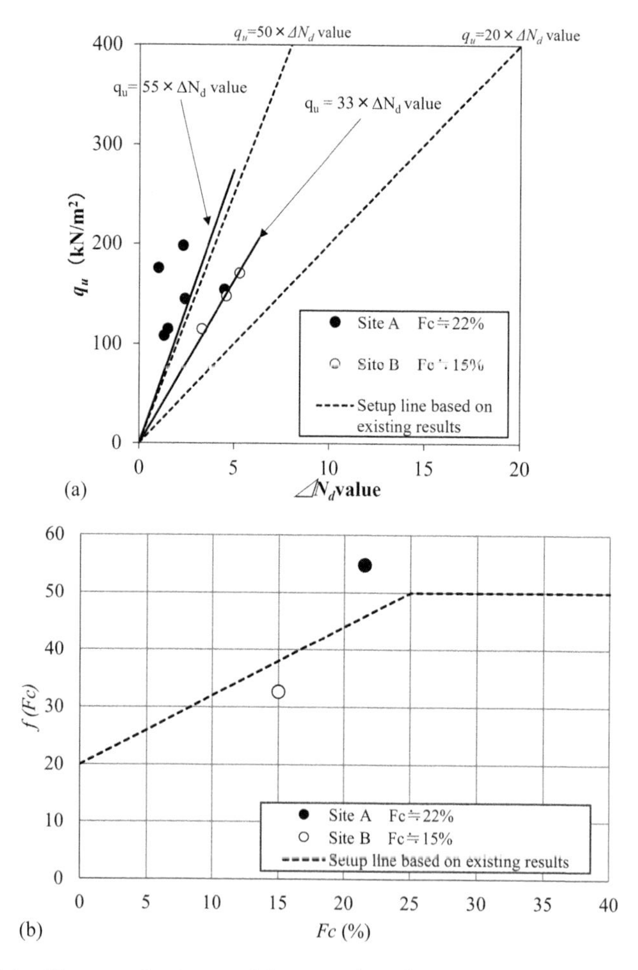

FIGURE 4.2 Test results in model ground and in the field. (a) Relationships between ΔN_d value and q_u. (b) Relationships between F_c and $f(F_c)$.

high correlation between the ΔN_d values and the unconfined compression strength q_u. Figure 4.2(b) shows that the relationship between ΔN_d and unconfined compression strength q_u varies depending on the fine fraction content F_c of the soil.

In Figure 4.2(a), the setting lines corresponding to 0% and 25% fine fraction content F_c based on the previous results (Ports and Harbours Bureau, Ministry of Land, Infrastructure, Transport and Tourism, 2018) are shown as dashed lines. In addition, the setting lines based on the results of Site A and Site B are shown as solid lines.

Using the relationship in Figure 4.2, the estimated unconfined compression strength q_u (hereinafter referred to as the estimated unconfined compression strength q_u) can be calculated from the ΔN_d value obtained from the PDC test conducted on the improved ground using Equation (4.1).

$$q_u = f\left(F_c\right) \times \Delta N_d \tag{4.1}$$

where,

q_u: Estimated unconfined compression strength (kN/m^2)
ΔN_d: *Nd*-value increment before and after improvement
$f(F_c)$: Coefficient set from fine fraction content F_c.

Note that $f(F_c)$ in Equation (4.1) is shown in Equations (4.2) and (4.3).

$$\text{If } F_c < 25\% \quad f\left(F_c\right) = \left(20 + \left(F_c\right) \times 1.2\right) \tag{4.2}$$

$$\text{If } F_c \geq 25\% \quad f\left(F_c\right) = 50 \tag{4.3}$$

However, it should be noted that Equation (4.1) is not applicable when the soil properties (fine fraction content F_c) are different before and after the improvement. The method of calculating the estimated unconfined compression strength q_u using Equation (4.1) allows the estimated unconfined compression strength q_u to be calculated only from the PDC test results before and after improvement. The estimated unconfined compression strength q_u can be determined continuously in the depth direction within the range where PDC tests can be performed.

4.2.1.2 *Estimation of Unconfined Compression Strength q_u by Borehole Load Test*

The horizontal borehole loading test is a test method to determine the deformation modulus, *E*, etc., of the ground by pressurizing the borehole wall in the borehole using gas or hydraulic pressure and determining the relationship between the pressure and the displacement of the borehole wall (Japanese Geotechnical Society, 2013). It is known that the deformation modulus E obtained from the horizontal borehole loading test and the deformation modulus E_{50} obtained from the results of unconfined or triaxial compression tests are almost the same regardless of the soil material (Japanese Geotechnical Society, 2013).

In practice, in the specifications for road bridges, the two are designed to be equal with respect to the modulus of elasticity E of the ground when determining the horizontal coefficient of subgrade reaction of the pile (Japan Road Association, 2017).

This is because the strain levels in both tests are in approximately the same range. However, chemical-grouted sand also has a relationship between the unconfined compression strength q_u and the deformation modulus E_{50} in the unconfined compression test, as expressed in Equation (4.4).

$$q_u = E_{50} / \beta_{\text{LAB}} \tag{4.4}$$

where,

β_{LAB}: coefficient obtained from the unconfined compression test results of the indoor mixing test

Figure 4.3 shows the relationship between the deformation modulus E_{50} and the unconfined compression strength q_u based on the results of the repeated loading borehole load test and the unconfined compression

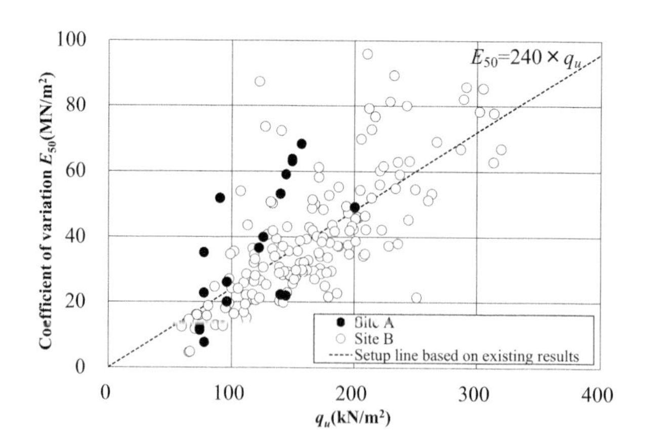

FIGURE 4.3 Relationships between PDC test results and q_u.

test using block sampling samples (Sugano et al., 2020) for the improved ground at Site B, with the case of Site A added to the diagram. The dashed line in Figure 4.3 shows the set line of the relationship between the deformation modulus E_{50} and unconfined compression strength q_u based on the previous results (Sugano et al., 2020) $E_{50} = 240 \times q_u$.

The case of Site A is based on the deformation modulus, E_{50}, from the borehole load test conducted on the improved ground, and the unconfined compression strength q_u, which is less affected by disturbance, obtained from the PDC test and the test a loading test using a triaxial test apparatus and silica content test described below. As shown in Figure 4.3, a correlation was confirmed between the two.

Using this relationship, the unconfined compression strength q_u (hereinafter referred to as the estimated unconfined compression strength q_u) can be calculated from the deformation modulus E_{50} obtained from the borehole load test in the improved ground using the following formula.

$$q_u = 240/E_{50} \tag{4.5}$$

where,

q_u: Estimated unconfined compression strength (kN/m^2)
E_{50}: Deformation modulus (kN/m^2)

However, the application of Equation (4.5) assumes that all pore spaces in the improved ground are filled with the chemical solution. The borehole load test is an in-situ test, which prevents the influence of disturbance at the time of sampling.

4.2.2 Strength Evaluation by Indoor Tests

4.2.2.1 Strength Evaluation by Triaxial Test Apparatus

Unconfined compression tests are also widely used to investigate strength properties of improved soils because of the simplicity of the test method compared to triaxial compression tests. When the subject of unconfined compression testing is a clay with low permeability, the unconfined compression strength q_u obtained has a clear mechanical implication that it is the shear strength at zero confining pressure under undrained conditions.

However, it is known that undrained conditions are not always ensured in shear in unconfined compression tests of treated sandy soils, and the strength of such soils is significantly reduced (Zen et al., 1990). Therefore, in order to ensure undrained conditions in unconfined compression tests

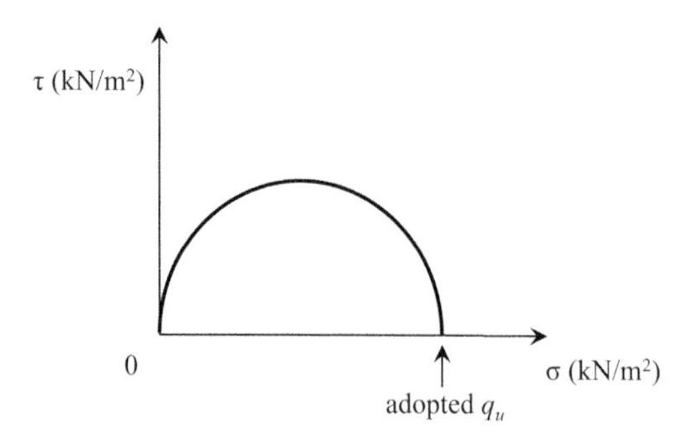

FIGURE 4.4 Method of checking q_u using triaxial test apparatus.

on sandy soils such as improved ground, a loading test using a triaxial test apparatus is proposed to ensure undrained conditions as in conventional unconfined compression tests, with reference to previous literature (Zen et al., 1990).

In this test method, a sampled specimen is placed in a triaxial test apparatus, and the specimen and duct are saturated by allowing water to flow through the specimen with a rubber sleeve attached. Then, as shown in Figure 4.4, a loading test is conducted under the condition that the undrained condition in the shear process is secured without any side pressure and without any consolidation process, and the unconfined compression strength q_u is obtained from the axial differential stress.

It should be noted that the results of loading tests using the proposed triaxial test apparatus should be checked for any disturbance of the specimens, such as the absence of a clear peak intensity from the stress-strain curve.

4.2.2.2 Estimation of Unconfined Compression Strength q_u by Silica Content Test

In the silica content test, silica gel packed in the pores of soil particles is eluted with a high concentration of alkali, and the eluted amount is measured. There are two types of silica content tests: ICP optical emission spectrometry (Coastal Development Institute of Technology, 2020) and atomic absorption spectrometry (Coastal Development Institute of Technology, 2020). Figure 4.5 shows the relationship between unconfined compression strength q_u and incremental silica content ΔS_i based on previous results (Coastal Development Institute of Technology, 2020), with the case of Site A added.

In Figure 4.5, the solid line shows the set line of the relationship between the unconfined compression strength q_u and the increment of silica

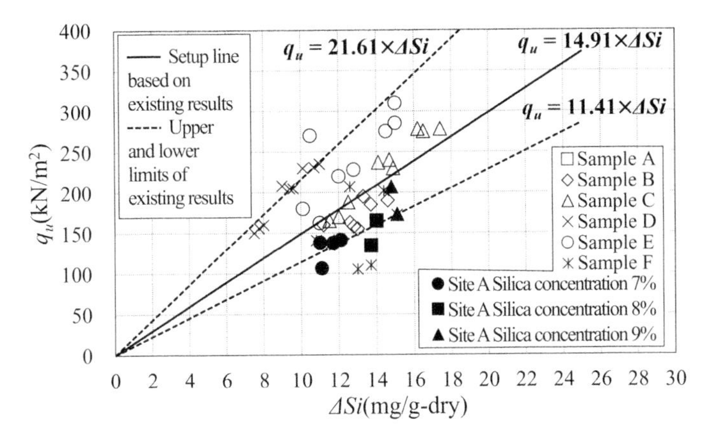

FIGURE 4.5 Relationship between q_u and incremental silica content ΔS_i.

content ΔS_i based on previous results (specimens A through F), and the dashed lines show the upper and lower limits of said relationship (Coastal Development Institute of Technology, 2020). The case study for Site A is based on the results of silica content tests and unconfined compression tests on specimens with silica concentrations of 7%, 8%, and 9%, which were conducted as formulation tests for the chemical grouting in the subject area.

Figure 4.5 shows that there is a correlation between the two. Using this relationship, it is possible to calculate the unconfined compression strength q_u of the improved ground (hereinafter referred to as the estimated unconfined compression strength q_u) from the incremental silica content ΔS_i obtained from the silica content test for the samples before and after improvement at the target location using the following formula:

$$q_u = A \times \Delta S_i \tag{4.6}$$

where,

q_u: estimated unconfined compression strength (kN/m²)

ΔS_i: incremental silica content of the improved ground relative to the remaining naturally occurring silica content in the pre-improved ground (mg/g-dry)

A: Proportionality coefficient of q_u and ΔS_i

However, since the silica content test is a physical test, Equation (4.6) can be considered an indirect strength estimation equation and should be applied only when other in-situ test or laboratory tests are not applicable. The coefficient A in Equation (4.6) is calculated from the relationship between the incremental silica content and unconfined compression strength q_u before and after mixing by conducting laboratory mixing tests and unconfined compression tests using samples taken at each site.

If the incremental silica content of the improved ground before and after the improvement can be determined, the estimated unconfined compression strength q_u can be calculated even if the specimens cannot be formed (Coastal Development Institute of Technology, 2020).

4.2.3 Correlation between Unconfined Compression Strength q_u and Estimated Unconfined Compression Strength q_u Based on the Results of Each Survey Method

Each of the survey methods described in (4.1) and (4.2) estimates the unconfined compression strength q_u, through different processes. Here, we evaluated the accuracy of estimating the unconfined compression strength q_u obtained by each method. To evaluate the estimation accuracy, the correlation coefficients between the unconfined compression strength q_u estimated from Equations (4.1), (4.5), and (4.6) and the unconfined compression strength q_u obtained from unconfined compression tests were checked for the data on which the above Equations (4.1), (4.5), and (4.6) are based.

The results of the correlation coefficients are shown in Figures 4.6–4.8. Figure 4.6 shows the relationship between the unconfined compression strength q_u (horizontal axis) from the unconfined compression test and the estimated unconfined compression strength q_u (vertical axis) from the ΔN_d value obtained by the PDC test. Correlation coefficients of 0.65 and 0.93 to 0.98 were obtained in the field test and indoor model test, respectively.

Figure 4.7 shows the relationship between the unconfined compression strength q_u (horizontal axis) from the unconfined compression test and the estimated unconfined compression strength q_u (vertical axis) from the deformation modulus E_{50}, with a correlation coefficient of 0.61. Figure 4.8 shows the relationship between the unconfined compression strength q_u (horizontal axis) obtained by the unconfined compression test and the estimated unconfined compression strength q_u (vertical axis) obtained by the silica content test using the increment of silica content ΔS_i. The correlation coefficient between the two is 0.40–0.98 for specimens A through F.

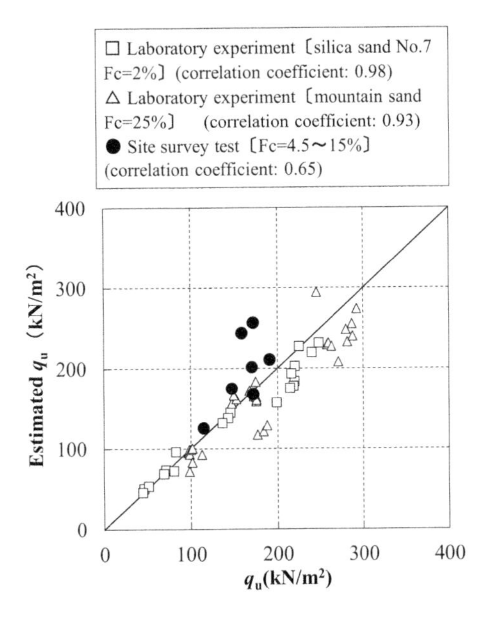

FIGURE 4.6 Relationship between q_u and estimated q_u. (Based on the data from the PDC that provide the basis for Equation (4.1)).

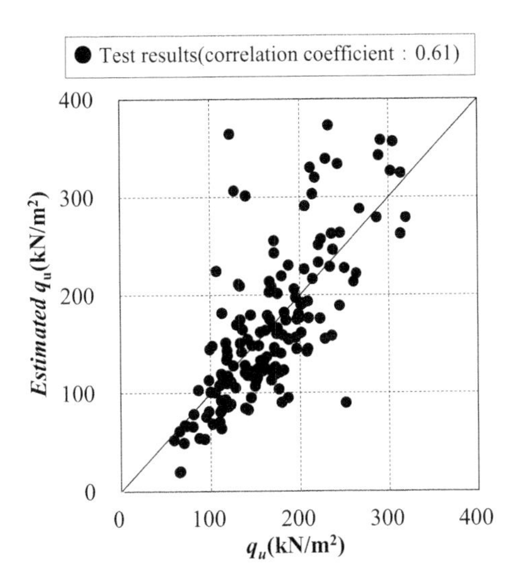

FIGURE 4.7 Relationship between q_u and estimated q_u. (Based on the data from the borehole load test that provide the basis for Equation (4.5)).

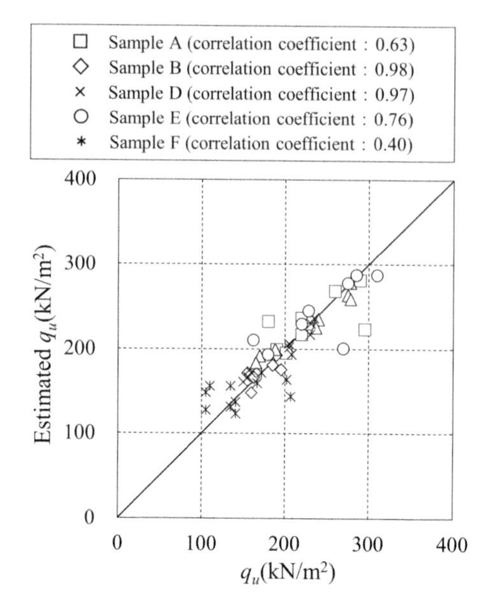

FIGURE 4.8 Relationship between q_u and estimated q_u. (Based on the data from the silica content test that provide the basis for Equation (4.6)).

As shown in Figures 4.6–4.8, the correlation coefficients were mainly approximately 0.6 to 0.7 for all the methods, except for the results of the laboratory model tests, which were conducted under ideal conditions. The correlation coefficients for the loading test using the triaxial test apparatus were not evaluated due to the lack of data compared to the results of the unconfined compression test.

However, in one example of laboratory mixing test, shown in Figure 4.1, the unconfined compression strength q_u from the unconfined compression test and loading test using the triaxial test apparatus shows relatively good correspondence, as well as the relationship between unconfined compression strength q_u and ΔN_d value, as shown in Figure 4.6.

The above results show that the accuracy of estimating the unconfined compression strength q_u of each survey method is not superior or inferior to the other. However, it is important to select an appropriate method based on the advantages, disadvantages, and application conditions of each method, and to use a combination of these methods according to the variation of soil properties at the target site (Table 4.1).

TABLE 4.1 Unconfined compression strength q_u by Loading Test Using Unconfined Compression Test and Triaxial Test Apparatus (Laboratory Mix Proportion Test at Site A)

		Unconfined compression strength q_u (kN/m²)		
Layer		**Unconfined Compression Test**		**Loading Test Using Triaxial Test Apparatus**
Alluvial sand (silica concentration 9%)	Sample *a*	206.95	197.6	120.0
		212.91		
		172.86		
	Sample *b*	83.98	126.0	
		151.48		
		137.07		
		131.43		

4.2.4 Strength Verification Methods and Application Conditions for Improved Ground

Table 4.2 summarizes the advantages and disadvantages of the above strength estimation methods using in-situ test and laboratory tests for improved ground. The methods in Table 4.2 can be classified into two groups: those that directly determine the unconfined compression strength q_u by laboratory tests and those that indirectly estimate the unconfined compression strength q_u based on the correlation between the measured values and the unconfined compression strength q_u based on previous knowledge.

For PDC testing, it is desirable that the pre-construction and post-construction surveys are conducted as close as possible to each other. It is not applicable when the ground conditions of the two are not consistent (negative ΔN_d or a difference of about 10% in the F_c values of the two).

The borehole load test cannot be conducted if the test depth interval is not more than 1.35 m (1.5 times the probe length of 0.90 m) or if the borehole wall is not easily self-supporting.

Strength estimation by silica content test is not applicable when there is no increase in silica content before and after improvement. As mentioned above, unlike other investigation methods, the silica content test estimates the unconfined compression strength q_u from the results of physical tests, and should be applied only when other in-situ test or laboratory tests are not applicable (Table 4.2).

TABLE 4.2 Strengths and Weaknesses of Each Survey Method

	In-Situ Testing Methods		Laboratory Test (Mechanical Test)			Laboratory Test (Physical Test)
Items	PDC[a]	Borehole Load Test[a]	Unconfined Compression Test[b]	Loading Test Using Triaxial Test Apparatus[b]	Cyclic Undrained Triaxial Test	Silica Content Test[a]
Research and testing method	• This is a dynamic cone penetration test using a special tip cone with a built-in pore water pressure meter to evaluate the N_d-value (equivalent to *N*-value) from the amount of penetration per blow, and to estimate the fine fraction content F_c from the excess pore water pressure in the ground generated by the blow penetration.	• In this test, the borehole wall is pressurized in the borehole using gas or hydraulic pressure, and the relationship between the pressure and the displacement of the borehole wall is used to determine the mechanical properties of the ground, such as the coefficient of deformation of the ground.	• This is a test to determine the maximum value of compressive stress (unconfined compression strength) by compressing a freestanding specimen in the longitudinal direction without any confining pressure.	• This is a test to determine the strength and deformation characteristics of soil when subjected to axial compression in an unconsolidated undrained condition.	• This is a test to determine the relationship between the one-amplitude or cyclic stress amplitude ratio of cyclic axial differential stress in undrained condition, the prescribed bi-amplitude axial strain, and the number of cyclic loadings until the prescribed excess pore pressure is reached in a saturated specimen compacted under isotropic stress conditions.	• Silica gel packed in soil particle pores is eluted with a high concentration of alkali.

How to determine liquefaction strength	• From the difference in N_d-values before and after the improvement (ΔN_d), the q_u value is obtained from the $\Delta N_d \sim q_u$ relationship.	• The q_u value is determined from the q_u relationship to the deformation coefficient E_{50} obtained from the borehole load test. • From the q_u to R_{L20} relationship, determines the liquefaction strength (R_{L20}).	• Based on the q_u value obtained from the test, the liquefaction strength (R_{L20}) is obtained from the q_u to R_{L20} relationship.	• Determine the q_u value from deviator stress at zero lateral pressure. • Based on the q_u value obtained from the test, the liquefaction strength (R_{L20}) is obtained from the q_u to R_{L20} relationship.	• The liquefaction strength (R_{L20}) is the ratio of cyclic stress amplitude at 20 cycles of cyclic loading in the liquefaction strength curve for both amplitude axial strains (DA=5%) obtained from the test. • If the number of specimens is not available, check the acceptability of a DA5% repetition rate Nc of 20 or more for the shear stress at design.	• The q_u value is obtained from the q_u relationship to the silica content increment obtained from the formulation test. • For the silica content test, the same test method shall be used before and after the improvement in order to prevent errors in test results due to differences in test methods. • From the q_u to R_{L20} relationship, determine the liquefaction strength (R_{L20}).

(*Continued*)

TABLE 4.2 (CONTINUED) Strengths and Weaknesses of Each Survey Method

	In-Situ Testing Methods		Laboratory Test (Mechanical Test)			Laboratory Test (Physical Test)
Items	PDC[a]	Borehole Load Test[a]	Unconfined Compression Test[b]	Loading Test Using Triaxial Test Apparatus[b]	Cyclic Undrained Triaxial Test	Silica Content Test[a]
Strengths	• Since it is an in-situ test, there are no effects of disturbance during undisturbed sample collection and trimming disturbance during the test. • Continuous data acquisition is possible.	• Since it is an in-situ test, there are no effects of disturbance during undisturbed sample collection and trimming disturbance during the test. • No data on the pre-improved ground is required (evaluation is possible with only information on the post-improved ground).	• q_u values can be obtained directly and simply. • No data on the pre-improved ground is required (evaluation is possible with only information on the post-improved ground).	• Less susceptible to disturbance than unconfined compression tests because of the applied confining pressure. • No data on the pre-improved ground is required (evaluation is possible with only information on the post-improved ground).	• The liquefaction strength can be determined directly. • No data on the pre-improved ground is required (evaluation is possible with only information on the post-improved ground).	• Since the test can be performed with a small amount of disturbed sample, relatively continuous data acquisition is possible.

Weaknesses	• Cannot be evaluated if the soil type is different before and after the improvement. • If the locations are too far apart before and after the improvement, evaluation may be difficult. • Evaluation may be difficult in soils with high N_d-values prior to improvement.	• Since it is necessary to maintain a test depth interval of at least 1.35 m (including the influence zone), it is not possible to determine the detailed improvement distribution status. • It is necessary to confirm that the chemical has penetrated into the ground by a separate method such as PDC (to check for an increase in pore water pressure) or sample collection (to check for an increase in silica content). • May not be applicable in soils where borehole walls are not easily self-supporting (gravelly soils).	• Since this is a test in which no confining pressure is applied, it is most susceptible to disturbances during sample collection and trimming during the test. • Based on past cases, it is not applicable in ground with high gravel content (over 15%).	• It is affected by disturbances during sample collection and trimming during testing. Therefore, it cannot be applied when the influence of disturbance is large. • Large dilatancy in shear may result in excessive strength ratings.	• Since there is also loading on the extensional side, it is more susceptible to disturbances during sampling and trimming during the test than loading tests using triaxial test apparatus. Therefore, it is not applicable when the influence of turbulence is large. • When cyclic load is high, it is difficult to obtain accurate liquefaction strength due to tensile failure during the test.	• In contrast to in-situ and mechanical tests, the q_u values obtained are estimates based on the results of physical tests (silica content). • In soils with large variations in soil properties, such as reclaimed land, it may be difficult to evaluate the silica content before improvement because of the large variations in the silica content.

Notes:

[a] Investigation to indirectly determine the unconfined compression strength q_u (estimated unconfined compression strength q_u).

[b] Investigation to directly determine unconfined compression strength q_u.

4.3 EVALUATION OF THE ENTIRE AREA OF IMPROVED GROUND USING A STRENGTH EVALUATION FLOW IN COMBINATION WITH VARIOUS INVESTIGATION METHODS

This section describes a case study of Site A, where the strength of the improved ground was confirmed based on the aforementioned investigation methods. In Site A, the blocks a through m shown in Figure 4.9 were improved using the chemical grouting, and post-construction surveys were conducted at all 51 points.

Figures 4.10 and 4.11 show an example of the ground characteristics of Site A and an example of the improvement specifications for the chemical grouting. As shown in Figures 4.10 and 4.11, the alluvial sand layer, which

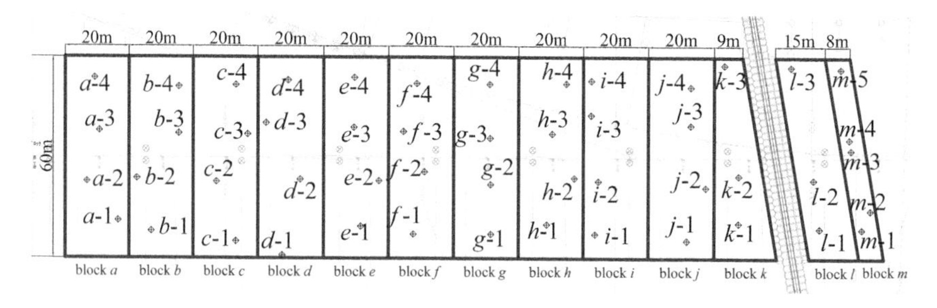

FIGURE 4.9 Range and plan of post-survey points with chemical grouting applied at Site A.

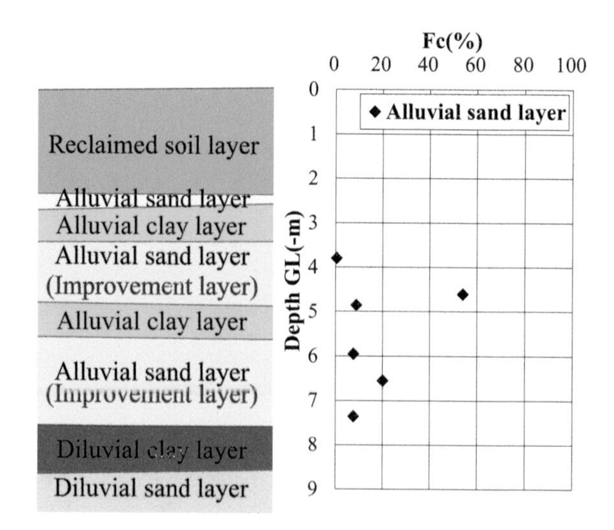

FIGURE 4.10 An example of the geotechnical characteristics of Site A.

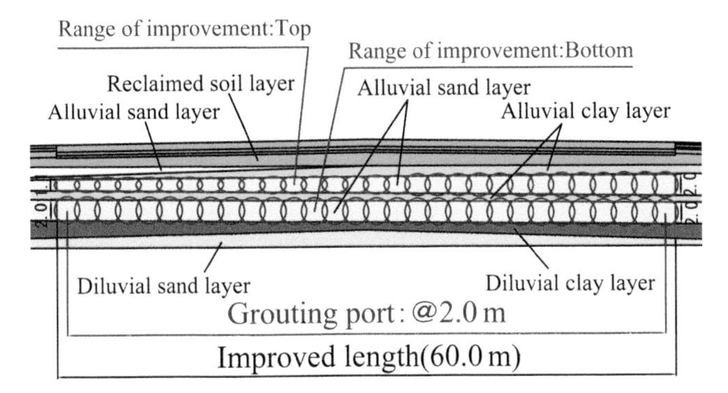

FIGURE 4.11 Example of improvement specifications for chemical grouting at Site A.

is the layer to be improved at Site A, is about 4 m thick, with a thin alluvial clay layer sandwiched within the layer. The fine fraction content F_c ranges from 0 to 20%.

The improvement was placed in a two-tiered arrangement with the upper and lower levels spaced 2.0 m apart, and the design standard strength was 60 kN/m^2.

The improved ground at Site A had a less complex stratigraphic structure and a lower gravel content than the subject soil. Therefore, it was considered possible to estimate the strength using PDC tests and borehole load test as in-situ tests and loading tests using a triaxial test apparatus as laboratory tests. Strength evaluations were conducted for the entire area of the improved ground based on the flow shown in Figure 4.12.

However, for the actual application, the unconfined compression strength q_u and the estimated unconfined compression strength q_u were continuously obtained at 20 cm intervals in the depth direction in order to understand the strength characteristics of the improved ground in detail. In the strength evaluation flow, first, as shown in Figure 4.12, the silica content test confirmed that the silica content increased compared to that before the improvement, thereby confirming the workmanship of the chemical grouting.

In addition to the incremental silica content, there is another method of confirming the workmanship using the excess pore water pressure ratio obtained from the PDC test (Coastal Development Institute of Technology, 2020). However, if the evaluation differs between the two test results, the silica content test results for the chemical component should be considered more important. Therefore, the verification of workmanship was based on the results of the silica content test.

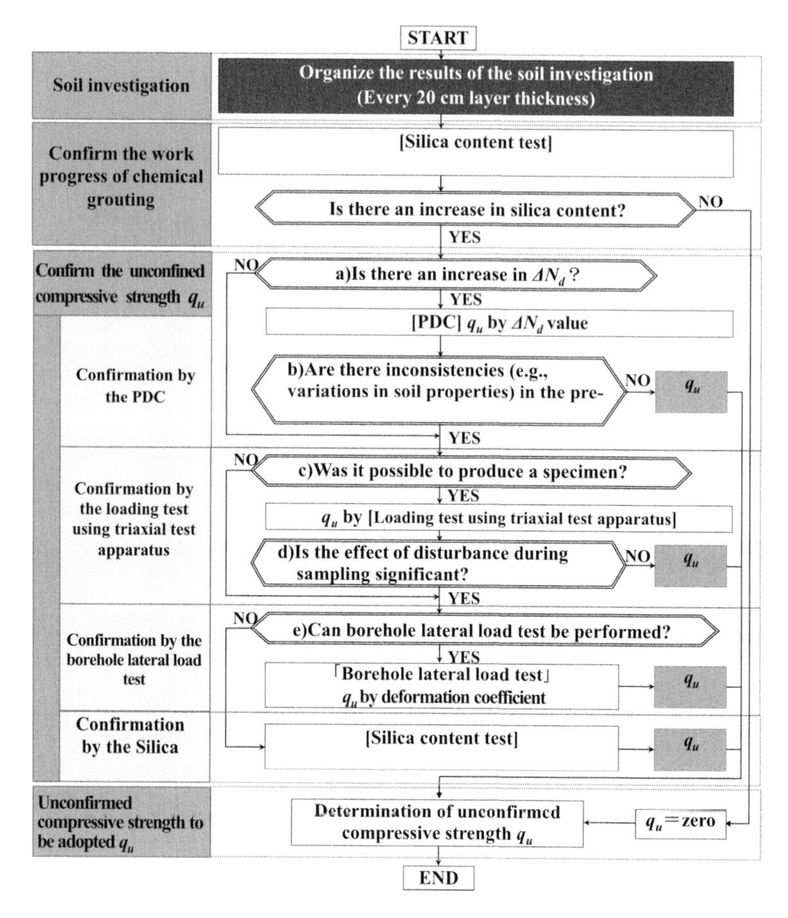

FIGURE 4.12 Flowchart for confirming the quality of workmanship, unconfined compression strength q_u, and estimated unconfined compression strength q_u at Site A.

In the subsequent strength evaluation, four different investigation methods were used together as needed. In Figure 4.12, the first PDC test was initiated. This is because the PDC test is an in-situ test without the influence of turbulence and has the feature of obtaining continuous data in the depth direction, which allows the quantity of investigation per investigation point to be increased for the layer to be improved at Site A, where the layer thickness is about 4 meters.

Next, if the N_d values did not increase before and after the improvement, or if the estimated unconfined compression strength q_u deviated from the range of previous results from laboratory mixing tests, even for depths where the N_d values increased, the N_d values were not adopted as the N_d values.

For depths where the PDC test results could not be employed, less disturbed specimens were taken, and when specimen compaction was possible, the unconfined compression strength q_u was directly measured by loading tests using a triaxial test apparatus. When specimen compaction was not possible, the unconfined compression strength q_u was estimated using a borehole load test. When it was difficult to perform any of the above tests, strength estimation was performed by silica content testing.

First, the increment of silica content in the ground before and after the improvement of block a as a representative block was checked, and the results of the workmanship check are shown in Figure 4.13.

The silica content before and after the improvement generally increased over the entire improvement area, confirming that the chemicals were injected in the improved area.

Next, the results of PDC tests on the ground before and after improvement were compared, and the estimated unconfined compression strength q_u was calculated from the fine fraction content F_c and Equation (4.1) in the subject area, as shown in Figure 4.14. As shown in Figure 4.14, the comparison of N_d and ΔN_d values before and after the improvement showed an increase in ΔN_d values after the improvement in the improvement range.

Figure 4.15 shows an example of the calculation results of unconfined compression strength q_u and estimated unconfined compression strength

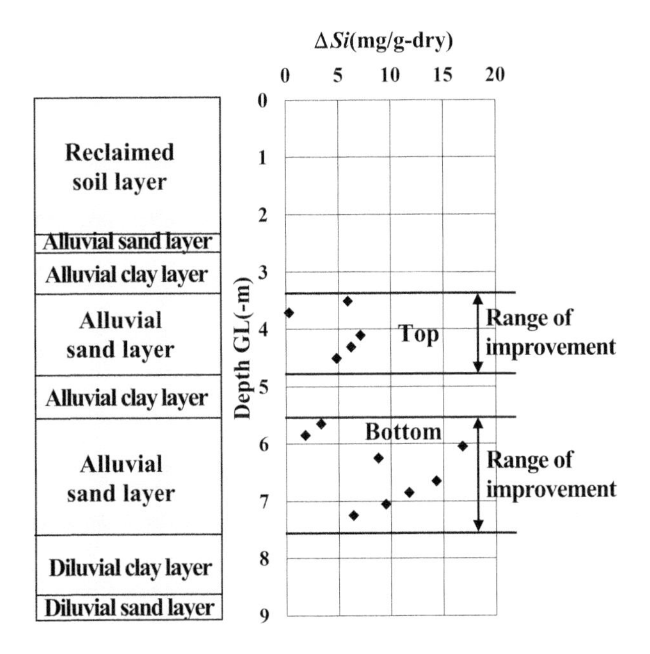

FIGURE 4.13 An example of the results of quality confirmation of workmanship.

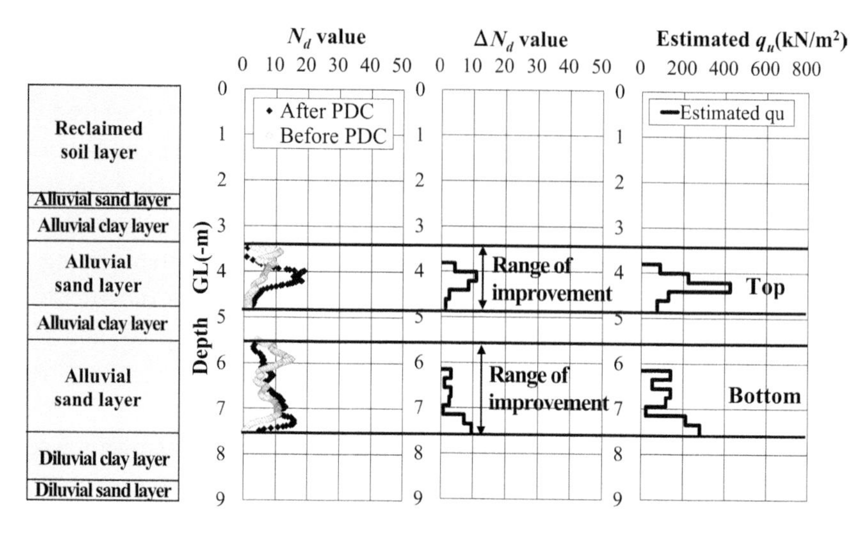

FIGURE 4.14 An example of the results of confirming the estimated unconfined compression strength (PDC test).

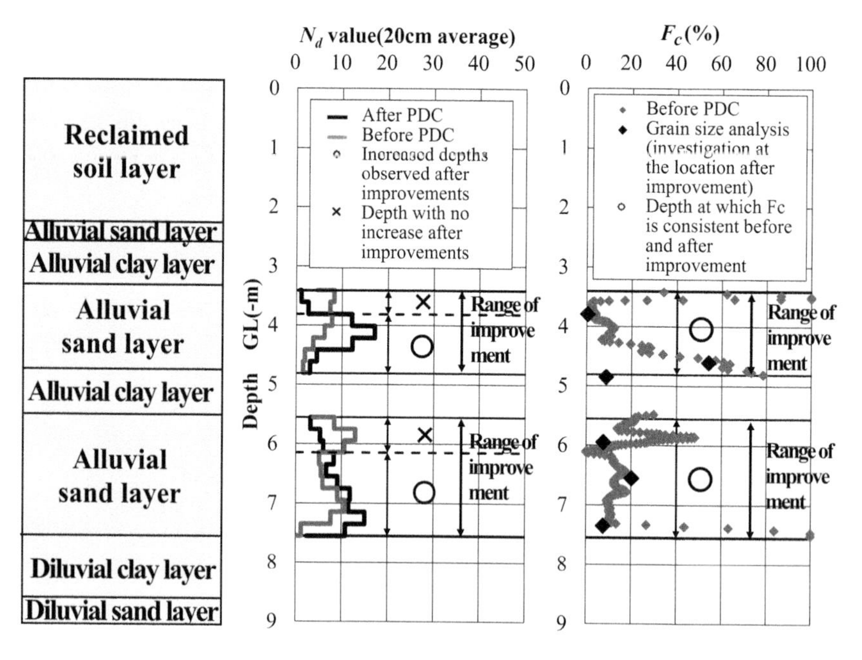

FIGURE 4.15 An example of the result of confirming the estimated unconfined compression strength q_u (an example of evaluating whether the PDC test result can be adopted or not).

q_u for each 20 cm layer thickness as an evaluation example of the PDC test results based on the flow shown in Figure 4.12.

The estimated unconfined compression strength q_u and the unconfined compression strength q_u obtained from the results of each test are listed in Table 4.3.

As shown in Table 4.3, the results of the workmanship verification of the chemical grouting showed that the silica content increased except for one layer (approximately 40 to 60 cm above the ground surface) where a silica content test was conducted, confirming that the workmanship of the method was secured. Next, the unconfined compression strength q_u and the estimated unconfined compression strength q_u were calculated for the layers for which the geometry was confirmed.

The ΔN_d values in each layer obtained from the PDC test were measured. As shown in Figure 4.15, the ΔN_d values increased in many layers, and the estimated unconfined compression strength q_u, could be calculated. However, there was no increase in ΔN_d values at the five depths.

For the fine fraction content F_c, the estimated value based on the excess pore water pressure measured by the PDC test and the fine fraction content F_c by the grain-size test were consistent throughout. The slight difference in the fine fraction content F_c between the two is considered to be due to the distance between the survey locations before and after the improvement.

For the upper one of the five depths where no increase in ΔN_d value was observed, specimens were formed from samples taken at that depth according to the strength evaluation flow, and the unconfined compression strength q_u was measured by loading tests using a triaxial test apparatus.

However, since it was not possible to form the specimens at the lower three depths where no increase in ΔN_d values was observed, borehole load test and strength estimation using Equation (4.5) were conducted.

Furthermore, the borehole load test was judged to be infeasible at the upper one depth where no increase in ΔN_d value was observed because of borehole wall disturbance, and the unconfined compression strength q_u was estimated from the silica content test results using Equation (4.6). By combining the various in-situ test and laboratory tests, the estimated unconfined compression strength q_u and the unconfined compression strength q_u could be calculated for all layers at 20 cm intervals where the geometry was confirmed.

TABLE 4.3 An Example of the Results (List) of the Confirmation of Estimated Unconfined Compression Strength q_u and Unconfined Compression Strength q_u

		Confirm the Work Progress		Confirm the Estimated Unconfined Compression Strength q_u and Unconfined Compression Strength q_u										Adoption of the Estimated Unconfined Compression Strength q_u and Unconfined Compression Strength q_u	
				Estimated Unconfined Compression Strength q_u by ΔN_d Value of PDC			Unconfined Compression Strength q_u by the Loading Test Using Triaxial Test			Estimated Unconfined Compression Strength q_u by the Deformation Coefficient of Borehole Load Test			Estimated Unconfined Compression Strength q_u by ΔS_i of Silica Content Test		
Range of Improvement	Stratigraphic Classification	ΔS_i by the Silica Content Test (mg/g-dry)	Comprehensive Evaluation*[a]	ΔN_d value *[b]	Fc*[b]	q_u (kN/m²)[c]	Specimen Production*[b]	Disturbance*[b]	q_u (kN/m²)*[c]	Availability*[b]	Test Depth (*)*[e]	q_u (kN/m²)*[c]	q_u (kN/m²)*[c]	Adopted Test*[d]	q_u (kN/m²)
Top	Alluvial sand layer	5.9	OK	No	Yes	— ➜	No	No	— ➜	No	—	— ➜	23.4	Silica	23.4
		0.3		No	Yes	— ➜	Yes	Yes	38.5	—	—	—	—	Loading	38.5
		0.0	NG	—	—	—	—	—	—	—	—	—	—	—	0.0
		7.1	OK	Yes	Yes	223.2	—	—	—	—	—	—	—	PDC	223.2
		6.2		Yes	Yes	425.0	—	—	—	—	—	—	—	PDC	425.0
		4.8		Yes	Yes	130.0	—	—	—	—	—	—	—	PDC	130.0
		22.7		Yes	Yes	75.0	—	—	—	—	—	—	—	PDC	75.0
*Alluvial clay layer		—	—	—				—			—		—	—	

Bottom	Alluvial sand layer	3.3	OK	No	Yes	—	➜	No	—	—	➜	Yes	*	**147.0**	—	Borehole	**147.0**
		1.8		No	Yes	—	➜	No	—	—	➜	Yes	*	**147.0**	—	Borehole	**147.0**
		16.8		No	Yes	—	➜	No	—	—	➜	Yes	*	**147.0**	—	Borehole	**147.0**
		8.8		Yes	Yes	**141.2**		—	—	—		—	—	—	—	PDC	**141.2**
		20.6		Yes	Yes	**52.9**		—	—	—		—	—	—	—	PDC	**52.9**
		14.3		Yes	Yes	**141.2**		—	—	—		—	—	—	—	PDC	**141.2**
		11.7		Yes	Yes	**119.1**		—	—	—		—	—	—	—	PDC	**119.1**
		9.5		Yes	Yes	**23.3**		—	—	—		—	—	—	—	PDC	**23.3**
		6.4		Yes	Yes	**212.6**		—	—	—		—	—	—	—	PDC	**212.6**
		21.5		Yes	Yes	**279.6**		—	—	—		—	—	—	—	PDC	**279.6**

Notes:
[a] Alluvial clay layer
[b] OK: Workmanship is secured. NG: Workmanship is not secured.
[c] Yes: Adoptable, because it meets the requirements. No: Cannot be adopted because it does not meet the requirements.
[d] **Bold text**: Adoption of the estimated unconfined compression strength q_u and unconfined compression strength q_u.
[e] PDC: PDC test, Loading: Loading test using triaxial test, Borehole: Borehole load test, Silica: Silica content test.
* Depth covered by borehole load test (0.6 m thickness).

4.4 HETEROGENEITY OF THE IMPROVED GROUND

In order to investigate the spatial variability in shear strength of improved ground, the results of unconfined compression strength q_u in Site A, which was improved by chemical grouting for the purpose of liquefaction countermeasure, are summarized. A proportional relationship between the liquefaction strength ratio $R_{L20,5\%}$ and the unconfined compression strength q_u of soil improved by chemical grouting and deep mixing treatment has been confirmed, as shown in Figure 4.16: the unconfined compression strength q_u increases with a constant slope when the liquefaction strength ratio increases (Coastal Development Institute of Technology, 2020).

Therefore, in principle, quality control of improved soil for liquefaction countermeasures is evaluated by checking whether the unconfined compression strength q_u of the undisturbed sample taken satisfies the design standard strength q_{uck}. The unconfined compression strength q_u was also used in the same construction project. Figure 4.17 shows a plan view of the area where the chemical grouting was implemented at Site A.

The construction work was divided into 13 blocks and was carried out. Site A is located on a plain that forms an urban area. Site A is located on a flat area where hills and terraces are formed and a river flows between them.

In the target area, layer B, which is the old fill, is about 2 m above the upper layer, the clay layer with low permeability is about 1 m, and the loose

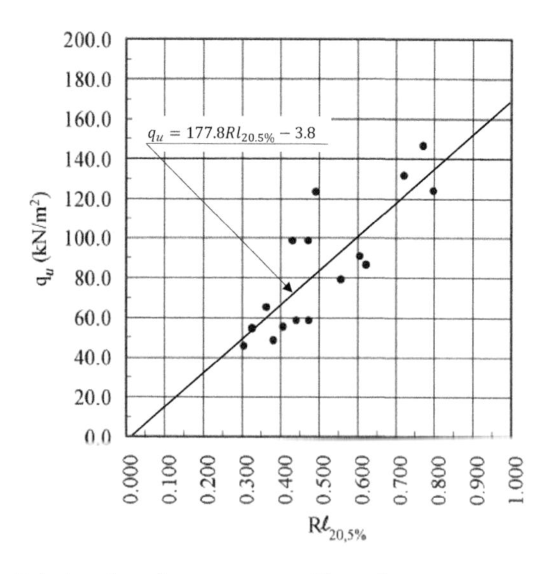

FIGURE 4.16 Relationships between q_u and liquefaction strength ratio.

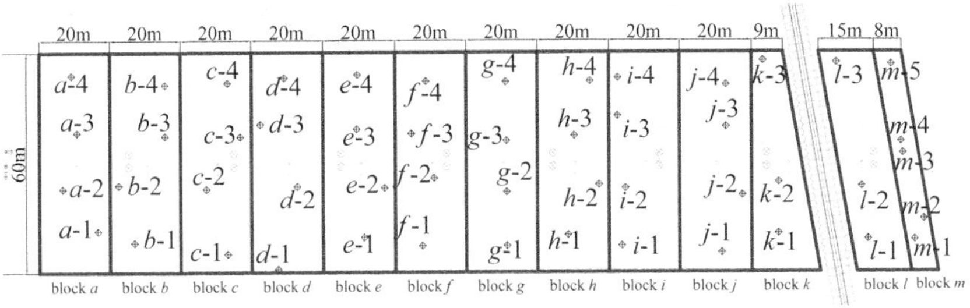

FIGURE 4.17 Range and block allocation with chemical grouting applied at Site A.

sandy soil layer is about 4 m below the upper layer. The sandy soil layer was determined to liquefy under the assumed L2 earthquake motion, and liquefaction countermeasures were implemented. An underground drainage pipe is located between Blocks k and l, and the ground has been improved by separate construction work.

The scope of the construction work to be organized is the range of blocks a through m, shown in Figure 4.17. The total improvement volume in the area where the measures were implemented was approximately 52,800 m^3, and a total of 51 points shown in Figure 4.17 were surveyed after the improvement construction work. A total of 897 data were obtained, with about 15 to 20 unconfined compression strength q_u data obtained at one site. The fine fraction content F_c of the soil to be improved within the construction area was generally about $F_c = 5–25\%$, porosity was $n = 39–44\%$, N-value was 3–15, and liquefaction strength R_{L20} was about 0.27–0.32.

A chemical grouting was adopted for this liquefiable layer, taking into consideration the impact on the existing runway pavement. The design base strength was $q_{\text{uck}} = 60$ kN/m^2, and chemical grouting was performed using silica at a chemical concentration of 9%. Figure 4.18 shows the frequency distribution of unconfined compression strength q_u in a representative block (h-block) at Site A.

The mean value of unconfined compression strength q_u was 125.9 kN/m^2 with a coefficient of variation of 0.544 for the 72 data samples.

The results show that the unconfined compression strength q_u of the improved ground is not necessarily the same value, and some spatial variability occurs.

Therefore, in the design of the chemical grouting, the in-situ surcharge factor ($\eta = 1.2$) is multiplied by the design base strength as a parameter to

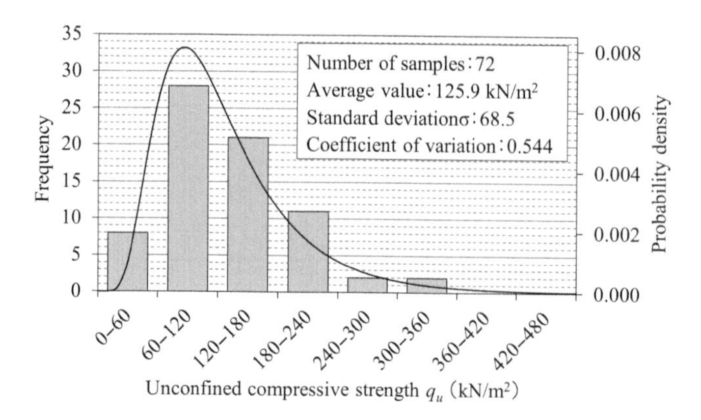

FIGURE 4.18 Distribution of unconfined compression strength q_u in a representative block at Site A.

account for the heterogeneity of the local soil. Since the distribution of the unconfined compression strength q_u tended to follow a log-normal distribution, the assumed log-normal distribution is shown as a solid line in Figure 4.18.

The causes of the spatial variability in unconfined compression strength q_u include the effects of disturbance during sampling and differences in the fine fraction content F_c and porosity of the source soil. The unconfined compression strength q_u summarized in this section includes some values from unconfined and triaxial UU tests conducted on triple-sampled specimens. However, most of the values were estimated from in-situ tests (N_d values from PDC tests, deformation modulus from borehole load test, etc.) and incremental silica content from silica content tests. Therefore, it is inferred that the effect of disturbance at the time of sampling is hardly included.

Despite the heterogeneity of the soil to be improved, the PDC test confirmed that the incremental excess pore water pressure and silica content of the improved ground ensured the workmanship of the soil.

Table 4.4 shows the results of goodness-of-fit tests for unconfined compression strength q_u and lognormal distribution.

Assuming a lognormal distribution for the unconfined compression strength q_u of the block in question, the chi-square value is 3.75, which is smaller than the variable value $C_{1-\alpha,f} = 12.6$ at the 0.05 level of significance and 6 degrees of freedom. It was found that the distribution of the measured unconfined compression strength q_u in the improved ground of the block conforms to a lognormal distribution.

TABLE 4.4 Goodness-of-Fit Test Results for Lognormal Distribution in Representative Blocks

Section of q_u (kN/m²)	Observed Frequency n_i	Theoretical Frequency e_i	$(n_i-e_i)^2/e_i$	Degree of Freedom
0~60	8	8.3	0.01	6
60~120	28	32.3	0.57	
120~180	21	19.2	0.16	
180~240	11	7.6	1.53	
240~300	2	2.8	0.24	
300~360	2	1.1	0.82	
360~420	0	0.4	0.42	
Total	72	72	3.75	

TABLE 4.5 Goodness-of-Fit Test Results for Lognormal Distribution in Each Block

Block	$(n_i-e_i)^2/e_i$
a	10.84
b	9.54
c	10.64
d	32.42
e	3.88
f	10.01
g	10.38
h	3.75
i	5.96
j	7.69
k	11.63
l	11.47
m	5.71

The unconfined compression strength q_u of each block obtained in the same way and the results of the goodness-of-fit test of the lognormal distribution are shown in Table 4.5.

This result gives a chi-square value of 32.42 for the d block, which is greater than the significance level of 0.05 and the variable value $C_{1-\alpha,f}$ = 12.6 with 6 degrees of freedom. The distribution of unconfined compression strength q_u for the same block did not fit a lognormal distribution. However, in the other blocks, the distribution of the measured unconfined compression strength q_u was found to fit a lognormal distribution.

Figure 4.19 shows the frequency distribution of unconfined compression strength q_u for 897 samples for all 13 blocks in the improvement range.

The average unconfined compression strength q_u was 137.1 kN/m^2 with a coefficient of variation of 0.623. Table 4.6 shows the results of goodness-of-fit tests for unconfined compression strength q_u and lognormal distribution.

Assuming a lognormal distribution for the unconfined compression strength q_u, the chi-square value is 4.58, which is smaller than the significance level of 0.05 and the variable value $C_{1-\alpha,f}$ = 12.6 with 6 degrees of freedom. From this, it can be said that the distribution of the measured unconfined compression strength q_u in the improved ground at Site A conforms to a log-normal distribution. This trend has also been observed in other construction projects shown in the *Technical Manual for Permeation*

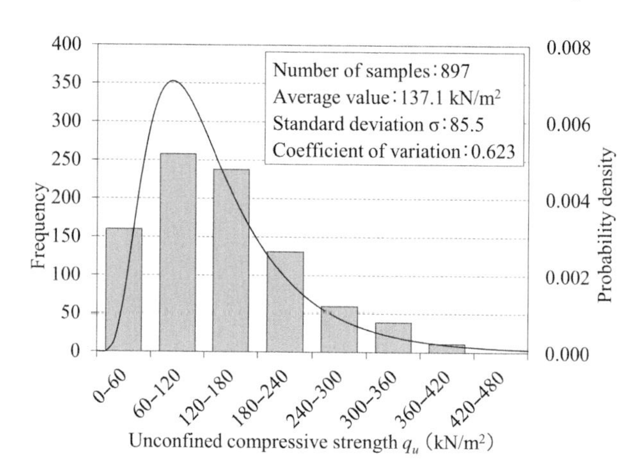

FIGURE 4.19 Distribution of q_u (kN/m^2) of samples collected at Site A.

TABLE 4.6 Goodness-of-Fit Test Results for Lognormal Distribution in Each Block

Section of q_u (kN/m^2)	Observed Frequency n_i	Theoretical Frequency e_i	$(n_i-e_i)^2/e_i$	Degree of Freedom
0–60	160	121.9	0.72	6
60–120	258	356.6	2.00	
120–180	238	229.0	0.01	
180–240	131	107.6	0.35	
240–300	60	48.5	0.46	
300–360	39	22.3	0.99	
360–420	11	10.6	0.05	
Total	897	896.5	4.58	

Grouting Method, with cases of coefficients of variation of unconfined compression strength q_u ranging from 0.2 to 0.5 and distribution of unconfined compression strength q_u based on post-facto investigations (Coastal Development Institute of Technology, 2020).

Therefore, the unconfined compression strength q_u of the ground properly improved by the chemical grouting is considered to fit a log-normal distribution.

To proceed with the statistical method, the unconfined compression strength q_u studied in each block was assumed to be log-normally distributed, and the goodness-of-fit was examined for all 13 blocks.

The compliance ratio is the ratio of the strength that satisfies the design basis strength to the population of the probability distribution obtained by assuming the distribution of shear strength to be a lognormal distribution.

Table 4.7 shows the calculation results of the compliance ratio for each block.

The results showed that the compliance rate in each block ranged from 62 to 97%.

The reasons for the differences in the compliance ratio and coefficient of variation for each block are thought to be that the distribution of fine fraction content F_c and porosity differs from block to block, and that some blocks have intervening areas with a large amount of gravel or clay content.

Unlike materials such as concrete, the heterogeneity of improved ground requires consideration not only of mean values and coefficients of

TABLE 4.7 The Percentage of Defective *PD* in Each Block

Block	Number of Boreholes	Data Number of q_u (kN/m^2)	Coefficient of Variation	The percent of *PD* (%)
a	4	86	0.590	91
b	4	82	0.774	77
c	4	83	0.552	94
d	4	81	0.654	86
e	4	65	0.846	62
f	4	68	0.638	83
g	4	68	0.540	94
h	4	74	0.544	89
i	4	78	0.714	80
j	4	73	0.541	95
k	3	58	0.586	92
l	3	52	0.466	97
m	3	48	0.616	83

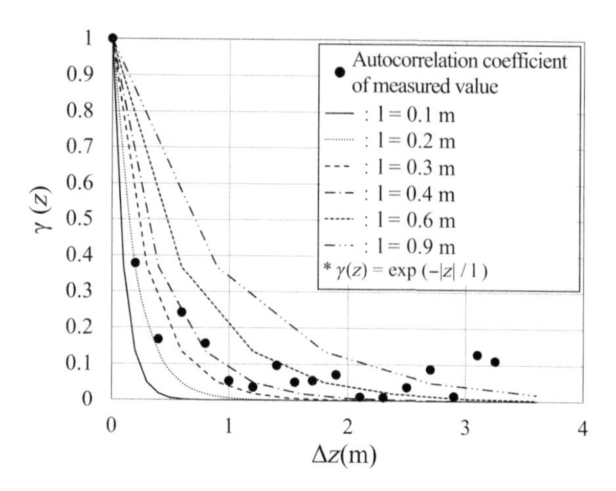

FIGURE 4.20 Autocorrelation coefficient of q_u (kN/m^2) in the vertical direction.

variation, but also of spatial autocorrelation using an autocorrelation function. Figure 4.20 shows the vertical spacing and autocorrelation coefficients for the unconfined compression strength q_u determined at 20 cm intervals in the same soil layer and with the same improvement specifications at a location in Site A.

The black plots in Figure 4.20 show the autocorrelation coefficient of unconfined compression strength q_u at intervals of approximately 20 cm from 0 to 325 cm.

Each line in Figure 4.20 is an exponential approximation of the autocorrelation coefficient at a pitch of 0.1 m to 0.3 m. Figure 4.20 shows that the vertical autocorrelation length is about 0.2 to 0.3 m, and that no autocorrelation is observed at greater distances.

As for the horizontal autocorrelation length, as shown in Figure 4.21, we confirmed the autocorrelation length with the survey points at 6 m, 24 m, 27 m, and 48 m, using the Bor. m–1 point as the base point.

Figure 4.22 shows the horizontal distance between points and the autocorrelation coefficient for the unconfined compression strength q_u between each point.

The black plots in Figure 4.22 show the autocorrelation coefficient of unconfined compression strength q_u for each distance between survey points, where each line is an exponential approximation of the autocorrelation coefficient at a pitch of 5 to 10 meters. Figure 4.22 shows that the horizontal autocorrelation length was scattered, resulting in almost no autocorrelation.

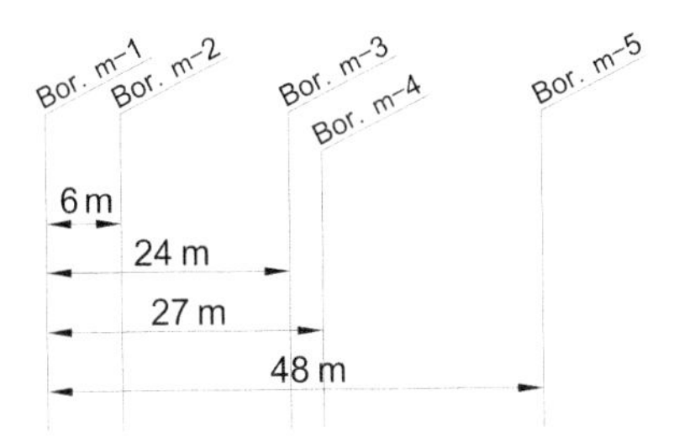

FIGURE 4.21 Survey location where horizontal autocorrelation length was checked.

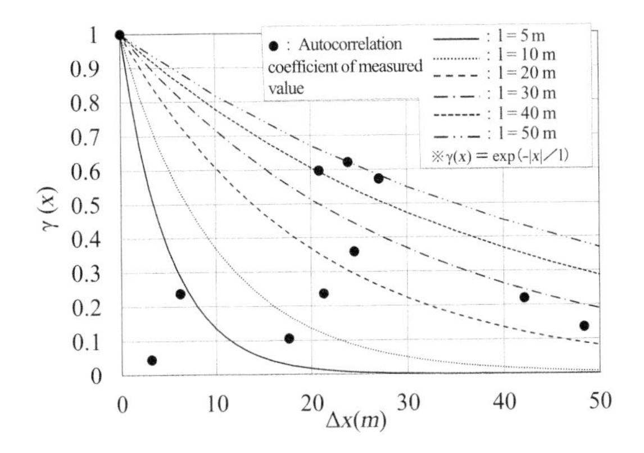

FIGURE 4.22 Autocorrelation coefficient of q_u (kN/m^2) in the horizontal direction.

These results confirm that the autocorrelation length of the improved ground strength is about 0.2 to 0.3 m in the vertical direction and about 2 m in the horizontal direction, indicating that the ground is nearly autocorrelation-free, i.e., random (Kasama et al., 2022).

4.5 CONCLUSIONS

In this chapter, a new strength verification method and heterogeneity of improved ground, which are less susceptible to sample disturbance, were studied to properly verify the strength of heterogeneous improved ground for liquefaction countermeasures in sandy soils. The results of the study revealed the following.

1) A method to estimate the unconfined compression strength q_u from the field test results before and after the improvement was proposed, utilizing the in-situ PDC test and the borehole load test in the borehole.

2) When in-situ testing is difficult, we proposed a method to calculate the unconfined compression strength q_u and estimated unconfined compression strength q_u by utilizing loading tests and silica content tests using a triaxial test apparatus, which are laboratory tests less susceptible to disturbance.

3) Points to note (application conditions), advantages and disadvantages of each strength evaluation method are summarized. The results of the chemical grouting test, which combined each strength evaluation method and the calculation of the estimated unconfined compression strength q_u and unconfined compression strength q_u of the improved ground, were also shown using data obtained from actual improved ground.

4) To evaluate the accuracy of the estimation of unconfined compression strength q_u by each of the investigation methods described in this section, the correlation coefficient between the unconfined compression strength q_u from the unconfined compression test and the estimated unconfined compression strength q_u by each method was calculated. As a result, we confirmed that the correlation coefficients for all survey methods were approximately 0.6 to 0.7. The accuracy of the estimation of unconfined compression strength q_u for each survey method was evaluated to be superior to that of the other methods.

5) The ground improved by chemical grouting had a spatial variability in shear strength due to the heterogeneity of the original soil before improvement, etc. The distribution of the measured unconfined compression strength q_u conformed to a logarithmic normal distribution. The autocorrelation length of the improvement strength for the same layer of soil with the same improvement specifications was 0.2 m to 0.3 m in the vertical direction and about 2 m in the horizontal direction, confirming that the soil is highly random.

REFERENCES

Coastal Development Institute of Technology: *Coastal Technology Library No. 55, Technical Manual for the Permeation Grouting Method* (Revised Edition), 2020.

Japan Road Association: *Specifications for Road and Bridge Part 4 Substructure*, 187–190, 2017.

Japanese Geotechnical Society: *Japanese Standards and Explanations of Geotechnical and Geoenvironmental Investigation Methods*, 2013.

Japanese Geotechnical Society: *Japanese Standards and Explanations of Laboratory Tests Geomaterials* (The 1st Revised Edition), 773, 2020.

Kasama, K., Nagayama, T., Hamaguchi, N., Sugimura, Y., Fujii, T., Kaneko, T. and Zen, K.: Performance-based evaluation for the bearing capacity of ground improved by permeation grouting method, *Journal of Japan Society of Civil Engineers Proceedings C (Geotechnical Engineering)*, Vol. 78, No. 1, 45–59, 2022.

Kawamura, K., Tanaka, K., Zen, K., Kasama, K. and Mine, N.: Evaluation method of improvement strength by in-situ tests of solution chemical injection methods, *57th JSCE Annual Meeting*, 203–204, 2002.

Okada, K., Kouzuki, K., Sasaki, T. and Suemasa, N.: The example of evaluation by the field test in foundation improvement by chemical injection of a quay in use, *Proceedings of the 10th National Symposium on Ground Improvement*, Japan, pp. 47–54, 2012.

Okada, K. and Takematsu, K.: Measures against liquefaction of ground by permanent grout—Multi-points grouting method—Challenges and countermeasures for ground hazards during earthquakes lessons learned and recommendations from the 2011 Great East Japan Earthquake (Secondary), *Japanese Geotechnical Society*, 304–305, 2012.

Ports and Harbours Bureau, Ministry of Land, Infrastructure, Transport and Tourism: Study on ground improvement work by injection grouting method at reclaimed land, 4th Committee Documents, 11, 2018.

Sugano, T., Zen, K., Suemasa, N., Kasugai, Y., Yamazaki, H., Hayashi, K., Sawada, S., Endo, T., Kato, T., Nakagawa, H., Kiku, H., Yamaguchi, E., Fujii, N., Baba, K., Fujii, T. and Takada, K.: Study on strength evaluation technique by using in-situ tests on chemical grouted ground as a countermeasure for liquefaction, Technical Note of the Port and Airport Research Institute, No. 1366, 2020.

Suwa, Y., Suemasa, N., Shimada, S. and Sasaki, T.: Properties of improved body components that influence strength prediction of chemical improved bodies, *Proceedings of the Japan National Conference on Geotechnical Engineering*, Vol. 41, 789–790, 2006.

The Ports and Harbours Association of Japan: *Technical Standards and Commentaries for Port and Harbour Facilities in Japan*, 786–787, 2018.

Yoshizawa, T., Sawada, S. and Nobumoto, M.: Study on the effect of blowing speed on the estimation of fine grain content by piezo drive cone, Japan Geotechnical Consultants Association, Geo Tech Forum 2013 in Nagano, No. 67, 2013.

Zen, K., Yamazaki, H. and Sato, Y.: Strength and deformation characteristics of cement treated sands used for premixing method, *Report of the Ports and Harbor Research Institute*, Vol. 29, No. 2, 111–113, 1990.

CHAPTER 5

Performance-Based Evaluation on the Bearing Capacity

Kiyonobu Kasama

5.1 INTRODUCTION

Japan is an earthquake-prone country, and in past earthquakes, liquefaction has caused extensive damage. The social infrastructure protecting the residents' lives and the economy must appropriately cope with seismic risks. In particular, transportation infrastructures such as ports and airports are required to maintain their functions even in the event of a large earthquake, because they act as transportation hubs in the event of a disaster. From this background, ground improvement techniques such as deep-mixing (DMM; Terashi and Tanaka, 1981) and pre-mixing (Zen et al., 1992) methods are becoming widely established for stabilizing soft soils in applications ranging from the strengthening of weak foundation soils to the mitigation of liquefaction. Although there have been significant advances in the equipment and methods used for ground improvement, several factors affect the undrained shear strength of the improved ground, including types and amounts of the binder/cement, physico-chemical properties of the in-situ soil, curing conditions, and effectiveness of the mixing process. Therefore, it is not surprising that the in-situ undrained shear strength of improved ground shows a large degree of spatial variability.

DOI: 10.1201/9781032670133-5

Kasama et al. (2012) reported that the mean values of unconfined compression strength q_u obtained from core samples cured in the field of improved ground range from 100–7500 kPa with coefficients of variation, $COV_{qu} = \sigma_{qu}/\mu_{qu} = 0.14–0.99$. These results are consistent with the findings of a review of US deep mixing projects by Navin and Filz (2005), who report $COV_{qu} = 0.17–0.67$. This level of variability is much higher than that expected for the undrained shear strength of natural clays. Kasama et al. (2012) found that correlation length can range from $\theta_h = 0.15–12.0$ m horizontally, and vertically $\theta_r = 0.2–4.0$ m (this is similar to the range of fluctuation scales quoted for natural clay deposits). Navin and Filz (2005) found that the horizontal correlation length is much larger for wet mix DMM block columns (12 m) than for dry-mix (<3 m). For the spatial variability in a single column, Larsson et al. (2005) discovered a radial correlation length, $\theta_r < 0.15$ m within 0.6 m diameter column.

Overall, these data suggest that the horizontal correlation length for improved ground is much smaller than for natural sedimentary soil layers, although some studies have found that the horizontal scale of fluctuation can be an order of magnitude greater than the vertical scale. Recently, Chen et al. (2016) and Liu et al. (2017 and 2019) investigated the interaction between the spatial variations in cement slurry concentration and in-situ water content in improved ground using statistical analysis and random field simulation based on field data. They reported that the scale of fluctuation of undrained shear strength of improved ground is much larger than a column diameter due to the variation of in-situ water content.

This spatial variability in the physical and mechanical properties of the improved ground (i.e., unit weight, undrained shear strength, etc.) introduces uncertainties in the design of foundations on improved ground. According to the *Technical Manual of the Deep Mixing Method for Ports and Airports*, if a relatively large amount of test result data is available, the strength distribution of the improved ground is considered as a normal distribution, and the strength of the improved ground can be evaluated using a statistical method to confirm the quality of this spatially variable improved ground. From the viewpoint of quality control and assurance for improved ground, the concept of percentage of defective (PD) samples has been introduced instead of the safety factor (SF). It is general for the design of ground improvement that the *SF*—the ratio of field strength depending on site condition variation (such as non-uniformities of mixing cement and curing condition) and required strength to stabilize the ground—is used to confirm the performance and quality of improved ground. Highly variable quality of improved ground not only increases the possibility of occurrence of defects during construction but also necessitates the use of a large *SF* in mix design to achieve the required strength, making the ground improvement uneconomical.

On the other hand, PD is defined as the percentage of defective samples having a quality level below the specific quality. Notably, PD is calculated as the volumetric ratio of strength that does not satisfy the design undrained shear strength value q_{uck} to the population of the probability distribution obtained assuming a normal distribution of field undrained shear strength q_u. For example, the concept of PD has already been introduced and applied for the quality control and assurance of ready-mixed concrete, and PD less than 5% is conventionally required in the quality control and assurance of concrete strength for the concrete engineering, as reported in JSCE (2007) and ACI (2011). In addition, the PD has been utilized for the quality control and assurance of ground improvement in Japan, and the PD up to ~15% can be sanctioned for the ground improvement of marine construction (The Coastal Development Institute of Technology, 2018). However, the appropriate range of *PF* for the quality control and assurance of ground improvement was not well discussed and summarized.

In order to evaluate the effects of the spatial variability of soil parameters on the bearing capacity problem, Griffiths and Fenton (2001), Griffiths et al. (2002), and Popescu et al. (2005) clarified the effects of the spatial variability of the material constants for the flat foundations on the cohesive ground using Monte Carlo simulations with conventional elastoplastic finite element analysis under plane strain conditions. For the bearing capacity of a single improved column, Namikawa and Koseki, 2013 and Namikawa (2016) investigated the effects of the spatial correlation of the undrained shear strength on the full-scale strength using similar random finite element analyses. Pan et al. (2018) investigated the effect of spatial variability on the short- and long-term strength of a single improved column against vertical loading in using an advanced effective stress constitutive model and random finite element analysis. Instead of random finite element analysis, Jellali et al. (2005) used a homogenization approach in three-dimensional limit analysis to model the geometry and strength for column-type improved ground. It is noted that in order to examine the effect of spatial variability for column-type improved ground consisting of an original soft ground with cylindrical columns of improved soil, three-dimensional analysis will be needed to represent the three-dimensional geometry and boundary condition on the bearing capacity of column-type improved ground.

Additionally, Kasama et al. (2012 and 2019) used Monte Carlo simulations with numerical limit analyses (Lyamin and Sloan, 2002; Sloan and Kleeman, 1995) to stochastically and statistically clarify the bearing capacity characteristics by expressing the spatial variability of the strength in random field theory for the ground bearing capacity analysis and proposed allowable percentage of defective for a simplified quality control and assurance in in-situ strength of full-replacement and a partial zone/block ground improvements, respectively. Although there have been significant advances in numerical methodology to consider the spatial variability on the bearing capacity problem of improved ground mentioned above, there are few prior studies to discuss and propose the appropriate range of *SF* and *PF* from the viewpoint of the quality control and assurance of ground improvement in terms of improvement pattern and dimension.

This study focuses on the ground improved by the chemical grouting, which is one of the liquefaction countermeasure methods for ground improvement through solidification. Among existing ground improvement technologies, chemical grouting is a new ground improvement technology that can be applied for ground below existing structures such as airport runways without removing existing structure. The chemical grouting uses colloidal silica as permeation grout, whose hardening time can be arbitrarily adjusted, to inject into the original ground through an injection pipe installed in the ground for the purpose of increasing the strength of the ground. Permeation grouts are injected into voids among soil particles without changing the structures of soil skeleton and solidify after hardening time. Injection of permeation grouts causes no damage of existing structures on/in the improved ground. The chemical grouting is, therefore, often employed for ground improvement, especially for liquefaction countermeasure, of port and airport in service.

Similar to other improved grounds, the infiltration variation of the chemical solution for chemical grouting—in addition to the original spatial variability of the soil properties of the concerned ground—produces a large spatial variability in the material constants compared to the naturally deposited ground. The spatial variability of the undrained shear strength improved by chemical grouting influences the bearing capacity evaluation of the improved ground for design. Hence, a bearing capacity evaluation method considering the spatial variability of the ground based on performance-based specifications is required for practical purposes. However, no previous studies have examined the strength spatial variability of improved ground based on the actual field data, and there are a few prior researches for a field application and case study using actual field data of proposed performance-based evaluation for bearing capacity.

In this chapter, the spatial variability of the undrained shear strength due to chemical grouting was carefully examined for the improved ground below the runway of an actual airport. The undrained shear strength was expressed in random field theory, and the bearing capacity analysis was conducted using Monte Carlo simulations with the finite element and shear strength reduction methods to account for the strength spatial variability and strength reduction due to liquefaction after an earthquake. Based on the results, the spatial variability effects due to chemical grouting on the bearing capacity safety factor, failure mechanism, and reliability of the bearing capacity were stochastically and statistically discussed.

Similar studies have been previously reported by Kasama et al. (2012 and 2019), who considered full-replacement ground improvement under vertical loading and a partial zone/block improvement of cement-treated ground beneath the footprint of the foundation under general loading conditions, respectively. This chapter describes a first application of proposed performance-based evaluation for an existing aircraft runway in Japan using field strength data. In addition, the size effect of the improvement area and the spatial variability effect of the undrained shear strength within the improvement area on the bearing capacity were also discussed after comparing the results of these studies with those of the present study, and then an appropriate range of percentage of defective (PD) for the quality control and assurance for ground improvement was finally proposed depending upon the ground improvement pattern and dimension.

5.2 BEARING CAPACITY ANALYSIS FOR CHEMICAL GROUTING IMPROVED GROUND

The program code SSR-FEM was used for the bearing capacity analysis (Zienkiewicz et al., 1975), which is an undrained shear strength reduction method based on FEM. The element type of the SSR-FEM is a four-node quadrilateral element wherein the Mohr-Coulomb failure criterion was applied with the associated flow rule as the plastic potential. The nodal reaction forces (residual forces) were evaluated from the element stresses in this elastoplastic calculation, and iterative calculations were performed for redistributing the residual forces until the solution converged.

The convergence of the iterative calculation was determined when the increment of all nodal displacements divided by the cumulative value of all nodal displacements was within 1.0×10^{-5} and the number of iterations was limited to 500.

The undrained shear strength was expressed using a Cholesky decomposition technique, and the FEM-based Monte Carlo simulation method with the shear strength reduction method was used for analyzing the bearing capacity considering the strength spatial variability. Since Kobayashi (1984) demonstrated that shear stiffness and Poisson's ratio pose marginal influences on the ultimate bearing capacity, it is assumed that shear stiffness and Poisson's ratio are constant in the bearing capacity analysis.

The effects of inherent spatial variability of soil property are represented in the analyses by modeling the undrained shear strength c of the improved ground soil as a homogeneous random field (Vanmarcke, 1984). The undrained shear strength is assumed to have an underlying log-normal distribution with mean, μ_c, and standard deviations, σ_c, and vertical and horizontal autocorrelation lengths, θ_v and θ_h. The current simulations assume that correlation length for unit weight, $\theta_{\ln\gamma}$, is similar to that for undrained shear strength, $\theta_{\ln c}$.

The mean and standard deviation of log c is readily derived from σ_c and μ_c as follows (e.g., Baecher and Christian, 2003):

$$\sigma_{\ln c} = \sqrt{\ln\left(1+\mathrm{COV}_c^2\right)} \tag{5.1a}$$

$$\mu_{\ln c} = \ln \mu_c - \frac{1}{2}\sigma_{\ln c}^2 \tag{5.1b}$$

Spatial variability is incorporated within the SSR-FEM by assigning the undrained shear strength corresponding to the ith element:

$$c_i = \exp\left(\mu_{\ln c} + \sigma_{\ln c} \cdot G_i\right) \tag{5.2}$$

where G_i is a random variable linked to the vertical and horizontal autocorrelation lengths, θ_v and θ_h.

Values of G_i are obtained using a Cholesky decomposition technique (CD, e.g., Matthies et al., 1997; Baecher and Christian, 2003; Kasama et al., 2006; Kasama and Whittle, 2011) using a Markov function, which assumes that the autocorrelation decreases exponentially with distance between two points i, j:

$$\rho_{ij} = \exp\left(-\frac{\sqrt{(x_i - x_j)^2}}{\theta_v} - \frac{\sqrt{(z_i - z_j)^2}}{\theta_h}\right) \tag{5.3}$$

where ρ_{ij} is the autocorrelation coefficient between element i and j, x and z are the directions of the vertical and horizontal coordinates, respectively, and $x_i - x_j$ and $z_i - z_j$ are the lag distances. Note that an exponential autocorrelation function is used to express the covariance structure of cement-improved ground, as experimentally shown by Honjo (1982), Navin and Filz (2005), and Larsson et al. (2005). It is emphasized that the coordinate at the centroid of the element in the finite element mesh for SSR-FEM is used to represent the spatial variability of undrained shear strength in this study. This coefficient can be used to generate a correlation matrix, $\boldsymbol{K}$, which represents the correlation coefficient between each of the elements used in the SSR-FEM finite element meshes:

$$\boldsymbol{K} = \begin{bmatrix} 1 & \rho_{12} & \cdots & \rho_{1n_e} \\ \rho_{12} & 1 & \cdots & \rho_{2n_e} \\ \vdots & \vdots & \ddots & \vdots \\ \rho_{1n_e} & \rho_{2n_e} & \cdots & 1 \end{bmatrix} \tag{5.4}$$

where ρ_{ij} is the correlation coefficient between element i and j, and n_e is the total number of elements in the mesh.

The matrix $\boldsymbol{K}$ is positive definite and hence the standard Cholesky decomposition algorithm can be used to factor the matrix into upper and lower triangular forms, $\mathbf{S}$ and $\mathbf{S}^{\mathrm{T}}$, respectively:

$$\mathbf{S}^T\mathbf{S} = \boldsymbol{K} \tag{5.5}$$

The components of $\mathbf{S}^{\mathrm{T}}$ are specific to a given finite element mesh and vertical and horizontal autocorrelation lengths, θ_v and θ_h.

The vector of correlated random variables, $\mathbf{G}$ (i.e., $\{G_1, G_2, \ldots, G_{ne}\}$, where G_i specifies the random component of the undrained shear strength in element i, Equation (5.2)), can then be obtained from the product:

$$\mathbf{G} = \mathbf{S}^T\boldsymbol{R} \tag{5.6}$$

where $\boldsymbol{R}$ is a vector of statistically independent random numbers $\{r_1, r_2,\ldots, r_{ne}\}$ with a standard normal distribution (i.e., with zero mean and unit standard deviation). Values of the random variable vector $\boldsymbol{R}$ are regenerated for each realization in a set of Monte Carlo simulations. A series of 100 Monte-Carlo simulations of SSR-FEM for a given analytical case has been performed. The detail of the analytical case is explained in the next section.

A generalized cross-section of improved ground at Airport A was modeled for the bearing capacity analysis. The improvement width was assumed to be the runway width of 60 m, while the improvement depth ranged from GL-4 m to GL-8 m (~4 m improvement thickness). In the analysis, the self-weight of the pavement on top of the ground was considered by uniform distributed loading. The finite element mesh is illustrated in Figure 5.1. The mesh spacing within the improvement area was 0.2 m × 0.2 m, considering the autocorrelation length of 0.2–0.3 m in the vertical direction for improved ground. The number of nodes and elements of the finite element mesh was set to 12,342 and 11,959, respectively. However, the area close to the aircraft loading uses fine mesh to ensure calculation accuracy. For the boundary conditions, the bottom displacements are fixed for the X- and Y-directions, and the side displacements are fixed for the X-direction but free for the Y-direction.

The aircraft load was assumed as B777-9, which is the largest aircraft among the LA-1 type aircrafts expected for takeoff and arrival to Airport A. The load characteristics of the B777-9 are presented in Table 5.1, and the wheel arrangement of the B777-9—including the loads used in the bearing

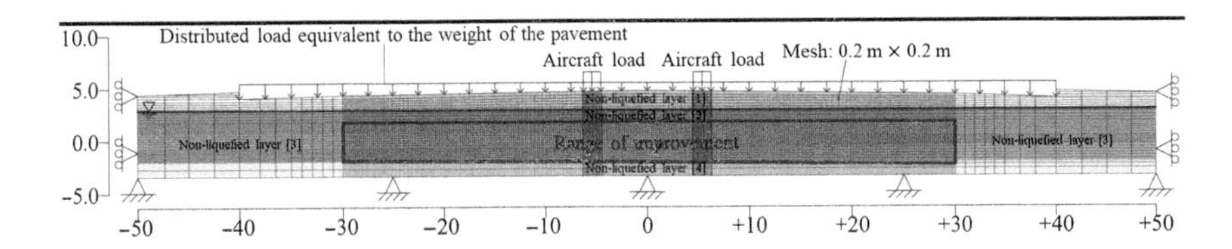

FIGURE 5.1 Numerical mesh.

TABLE 5.1 Load Specifications for B777-9

Aircraft	Total Mass (t)	Gear/Wheel Load (kN)	Ground Contact Pressure (N/mm²)	Ground Contact Width (cm)
B777-9	352.4	1630/272	1.58	34.4

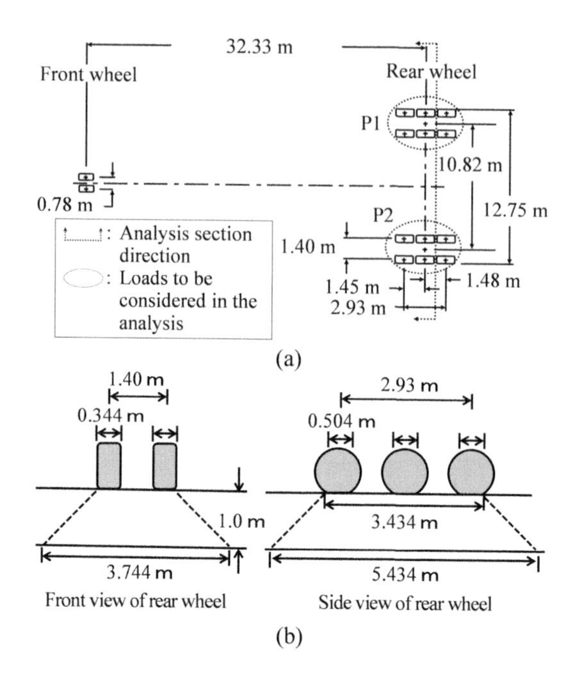

FIGURE 5.2 Wheel layout and contact load of B777-9 (a) Wheel layout, (b) Load distributed is considered by the analysis.

capacity analysis—is illustrated in Figure 5.2. As the bearing capacity analysis was performed in the transverse direction of the runway under the plane strain condition, the rear wheel loads, P1 and P2—as shown in Figure 5.2—were considered in the analysis. The aircraft loads were set in consideration of the load distribution on the pavement (GL-1 m). The aircraft load conditions are listed in Table 5.2.

The Civil Aviation Bureau, Ministry of Land, Infrastructure, Transport and Tourism (2019a and 2019b) stipulated a design standard for the

TABLE 5.2 Aircraft Load Condition

Aircraft		B777-9
Ground contact pressure		1.58 (N/mm^2)
Ground contact width		344 (mm)
Ground contact length		500 (mm)
Loads per a front wheel P_0		$P_0 = 1.58 \times 344 \times 500 \div 1000 \fallingdotseq 272$ (kN)
Aircraft loads from rear wheels (static aircraft load + accidental load as a 30% of static aircraft load)	P_1	$P_1 = 6 \times P_0/(1.400 + 0.344)/(2.930 + 0.504 + 1.000 \times 2) \times 1.3 \fallingdotseq 224$ (kN/m^2) ※ loading width on analysis model: 1.744 m
	P_2	$= P_1$

accidental load associated with aircraft traffic stating that it should be considered for underground structures below the runway when the depth of underground structure from the runway surface is less than 4 m. Since the depth to the improvement area of chemical grouting from the runway surface at Airport A was at most 3 m, which is less than 4 m, an additional load of the 30% static aircraft load was considered as an accidental aircraft load. The aircraft load is assumed to be the maximum load, while the load variation is not considered in the analysis.

The constitutive model of the improvement area and the original ground was assumed to be an elasto-perfect plastic with the Mohr-Coulomb failure criterion and the plastic potential with associated flow rule. The material constants included the soil unit weight γ, elastic modulus E_{50}, cohesive strength c, and Poisson's ratio v. The material constants for the constitutive model in the numerical analysis were determined based on the results of a site investigation at Airport A and laboratory test for soil samples obtained from Airport A, as shown in Table 5.3. The cohesive strength c ($=q_u/2$) of the improvement area was realized by a Cholesky decomposition technique to express the strength spatial variability in the assumed range of PD and coefficient of variation. Specifically, q_u of the improved ground was assumed as log-normally distributed, mean strength μ_{qu}, coefficient of variation COV, autocorrelation length θ, and normal random matrix X were used to calculate q_u of each element using the midpoint method (Matthies et al., 1997) with Cholesky decomposition (Baecher and Christian, 2003; Kasama et al., 2008).

For both vertical and horizontal autocorrelation lengths, Kasama et al. (2012) conducted a parametric study to investigate the effect of autocorrelation length on the bearing capacity of spatially random cohesive soil using numerical limit analyses and then concluded that the bearing capacity reduction becomes large with decreasing autocorrelation length. Therefore, the small vertical correlation length of 0.2 m was selected for improved ground from the range from 0.2 m to 0.3 m observed from measurements, as explained in Chapter 4. Additionally, it is generally known that the horizontal autocorrelation length of naturally deposited ground is larger than the vertical one, and it is also applicable for those of cement improved ground. Moreover, the horizontal autocorrelation length of improved ground was estimated less than 5.0 m.

Therefore, it is assumed in this study that the ratio between horizontal autocorrelation length and vertical autocorrelation length of improved ground was 10. Eventually, in this study the autocorrelation length of q_u

TABLE 5.3 List of Input Parameter

Layer	Soil	Unit Weight		N-value	Modulus of elasticity	Shear strength		Poisson
		Wet γ_t (kN/m^3)	Saturated γ_{sat} (kN/m^3)		E_{50} (kN/m^2)	c (kN/m^2)	ϕ (°)	ratio ν
Non-liquefied layer [1]	Sandy	18.9	19.4	11	30,500	8.3	37.7	0.33
Non-liquefied layer [2]	Clay	16.5	16.6	3	9500	18.5	18.4	0.33
Improvement area	Sandy	18.2	19.0	—	23,100	Set per case	0.0	0.33
Non-liquefied layer [3]	Sandy	18.2	19.0	8	21,000	20.3	35.3	0.33
Non-liquefied layer [4]	Clay	17.8	17.8	6	17,640	31.1	16.0	0.33

for the improved ground was assumed to be 0.2 m in the vertical direction and 2.0 m in the horizontal direction. Moreover, only the strength variations in the improved area caused by the chemical grouting were evaluated under two-dimensional plane strain conditions, without considering the variability in the original subgrade.

In order to calibrate the accuracy on the spatial variability of undrained shear strength by a Cholesky decomposition technique, as explained in Section 5.2, the q_u distribution was calculated for the mean q_u of 105 kPa and COV = 0.6. The histogram of q_u realized by a Cholesky decomposition technique is presented in Figure 5.3, and the typical example of q_u distribution in the finite element mesh is presented in Figure 5.4. It can be seen that mean q_u and COV obtained from a Cholesky decomposition technique coincide with those of input q_u and COV.

The bearing capacity analysis was performed under the following assumptions. 1) It is quite a rare case that an aircraft is in the motion of

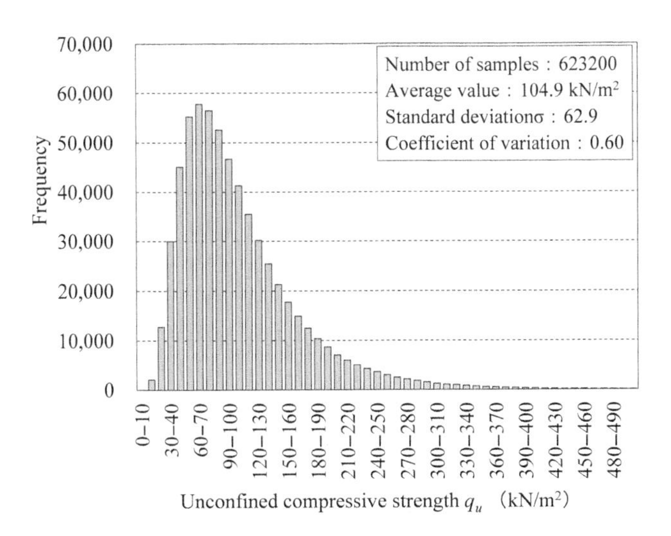

FIGURE 5.3 q_u distribution realized by random field theory.

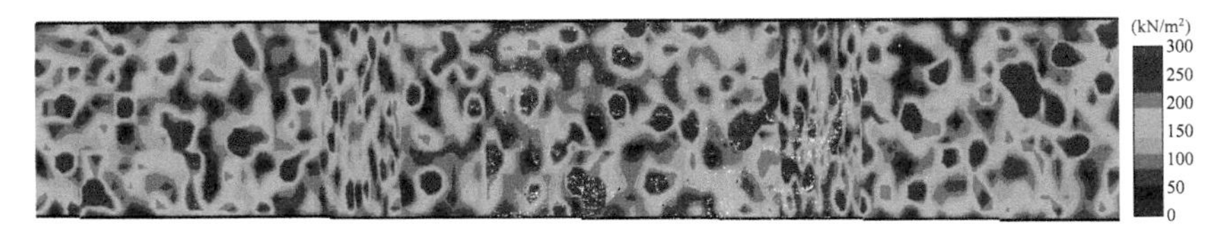

FIGURE 5.4 Example for strength spatial variability of q_u within improvement area.

takeoff or arrival on the runway when a large earthquake occurs. Therefore, the earthquake response analysis with/without aircraft load during an earthquake was not directly conducted in this study. In addition, the horizontal acceleration due to an earthquake was not considered in the bearing capacity analysis. 2) An important airport is required to maintain its functions even in the event of a large earthquake because it acts as a transportation hub in the event of a disaster. Specifically, in Japan, it is required that airport facilities reopen within three days after an earthquake. As a result, there is a possibility that excess pore water pressure buildup in airport ground occurs due to an earthquake and that it does not dissipate sufficiently within three days after the earthquake. Therefore, excess pore water pressure buildup due to the earthquake was considered in the bearing capacity analysis.

To evaluate from the safety considerations perspective, finite elements in the analysis that its q_u is less than q_{uck} were treated as if the liquefaction for the element had occurred during the earthquake, and the excess pore water pressure ratio for the element was constant at 1.0. It means that the undrained shear strength was set to zero for those elements whose q_u was less than q_{uck}. These assumptions were made because, in accordance with the performance requirements for major airports under a large earthquake, Airport A is expected to be used for the takeoff and arrival of aircraft for the purpose of transporting emergency supplies and personnel within three days after an earthquake.

As presented at Table 4.7 in Chapter 4, the coefficients of variation for q_u investigated for each block ranged from 0.466 to 0.846, while the PD for each block ranged from 3% to 38%. The current SSR-FEM calculation assumed that PD and COV of q_u are two main changing input parameters to investigate allowable PD irrespective of COV from the viewpoint of practical quality control and assurance for the bearing capacity of improved ground. Table 5.4 summarized input parameters for SSR-FEM. Case 1 is for the uniform q_u of 60 kN/m^2 without spatial variation of strength. Cases 2, 3, 5, and 6 are for combinations of PD (10%, 23.3%, 32%, 38.6%) and the COV of q_u (0.2, 0.4, 0.6, 0.8, 1.0). Case 4 is for a fixed PD = 28.0% and the COV of q_u = 0.2 and 1.0 to compensate for the result of bearing capacity safety factor in 20% range of PD. Cases 7 to 9 are for a fixed COV of q_u = 0.6 and PD = 51.1%, 61.7%, and 71.0% to investigate the bearing capacity safety factor for a very large PD. A total of 26 cases were conducted in this study. It is noted that the value of q_u in Table 5.4 was

TABLE 5.4 List of Analysis Cases

Case		Percentage of Defective (%)	Coefficient of Variation	Mean Strength (kN/m²)
1		0.0	—	60.0
2	*a*	10.0	0.2	73.9
	b	10.0	0.4	101.3
	c	10.0	0.6	136.1
	d	10.0	0.8	186.1
	e	10.0	1.0	237.3
3	*a*	23.3	0.2	58.9
	b	23.3	0.4	74.5
	c	23.3	0.6	94.4
	d	23.3	0.8	118.7
	e	23.3	1.0	148.6
4	*a*	28.0	0.2	68.4
	b	28.0	1.0	140.0
5	*a*	32.0	0.2	49.0
	b	32.0	0.4	62.0
	c	32.0	0.6	76.5
	d	32.0	0.8	91.5
	e	32.0	1.0	112.3
6	*a*	38.6	0.2	43.2
	b	38.6	0.4	55.7
	c	38.6	0.6	65.8
	d	38.6	0.8	76.9
	e	38.6	1.0	95.1
7		51.1	0.6	48.3
8		61.7	0.6	35.7
9		71.0	0.6	25.8

uniquely determined using the PD and COV under the assumption that q_u follows a log-normal distribution.

A series of 100 Monte Carlo simulations have been performed for the input parameters, as shown in Table 5.4. The computed bearing capacity safety factor, FS_i, can then be reported for each realization of the undrained shear strength field. The mean, μ_{FS}, and standard deviation, σ_{FS}, of the bearing capacity factor are recorded through each set of Monte Carlo simulations, as follows:

$$\mu_{FS} = \frac{1}{n}\sum_{i=1}^{n} FS_i ; \sigma_{\mathrm{FS}} = \sqrt{\frac{1}{n-1}\sum_{i=1}^{n}\left(FS_i - \mu_{FS}\right)^2} \tag{5.7}$$

5.3 BEARING CAPACITY CHARACTERISTICS CONSIDERING STRENGTH SPATIAL VARIABILITY

Initially, the results of Case 1 are presented, wherein the undrained shear strength was uniformly set to q_{uck} = 60 kN/m^2 such that q_u of all elements satisfies q_{uck} with no variability in the undrained shear strength (PD: 0%). In particular, the undrained shear strength was not varied, and thus, the bearing capacity safety factor was constant at 1.40. The maximum shear strain distribution and incremental displacement for Case 1 are presented in Figure 5.5, implying that the maximum shear strain extends up to the improvement range depending on the aircraft load; however, it is within the range of the improvement area. Moreover, the reflection and extension of shear strain are generated at the interface between the improved and non-liquefiable layer because of the stiffness difference between both layers.

Subsequently, a typical result obtained by varying the undrained shear strength within the improvement area by chemical grouting is presented in Case 3, where the PD was set to 23.3%. The results for Case 3a with COV = 0.2 and Case 3e with COV = 1.0 are mainly explained herein. The typical examples of strength distribution of q_u for Case 3a (PD: 23.3%, COV: 0.2) and Case 3e (PD: 23.3%, COV: 1.0) are plotted in Figure 5.6. The blue-colored areas in Figure 5.7 indicate elements that do not satisfy q_{uck} (elements with zero undrained shear strength in bearing capacity analysis).

The histogram of the bearing capacity safety factor obtained from 100 iterations for Cases 3a and 3e, same PD = 23.3% and COV = 0.2 and 1.0, respectively, is plotted in Figure 5.8. The mean safety bearing capacity safety factor for Case 3a, PD = 23.3% and COV = 0.2, was 1.23, which varied between 1.20 and 1.26. The bearing capacity safety factor in Case 3e, PD = 23.3% and COV = 1.0, varied between 1.35 and 1.49, which was

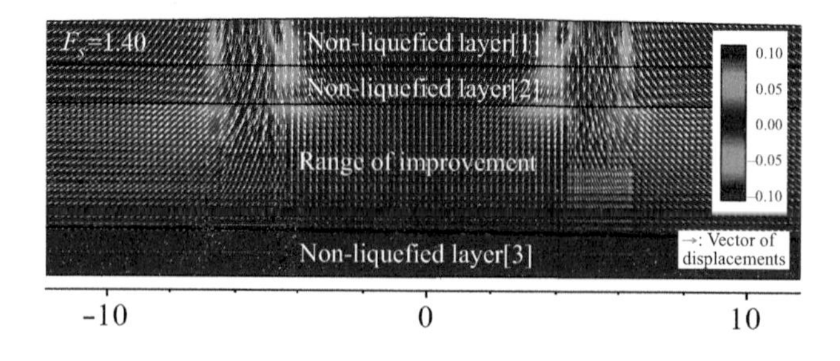

FIGURE 5.5 Maximum shear strain distribution for Case 1 (PD: 0%).

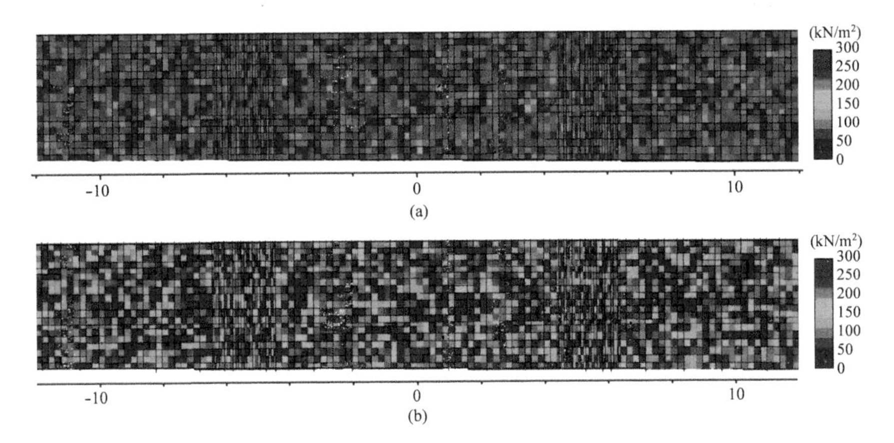

FIGURE 5.6 Typical example of strength distribution of q_u (a) Case 3a (PD: 23.3%, COV0.2), (b) Case 3e (PD: 23.3%, COV1.0).

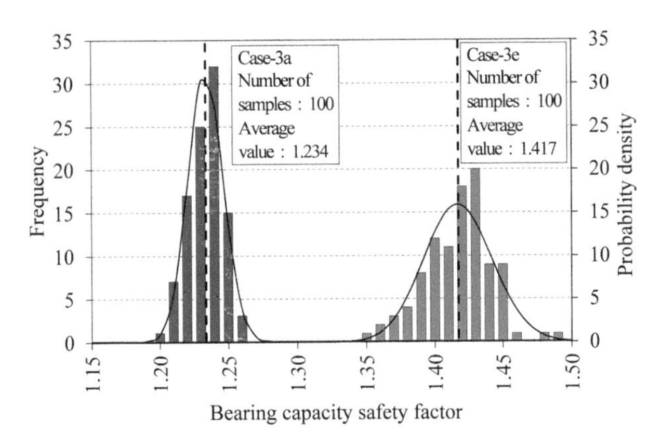

FIGURE 5.7 Histogram of the bearing capacity safety factor for Cases 3a and 3e.

expectedly larger than that of Case 3a (COV: 0.2). Overall, the mean value for Case 3e was 1.42, which was larger than that of Case 3a, while the coefficient of variation of bearing capacity factor for Case 3e was larger than that of Case 3a. Therefore, even though PD for both cases are the same values, mean bearing capacity and its variation are affected by mean and COV of shear strength of improved ground. It can be emphasized that PD is not a critical index on the evaluation of bearing capacity factor even though it is very simple to quantify the quality of improved ground.

To determine the number of Monte Carlo iterations required for the bearing capacity analysis, the mean and coefficient of variation of the bearing capacity safety factor against iteration number were summarized.

For instance, the mean bearing capacity safety factor against Monte Carlo iterations for Case 3 is presented in Figure 5.8, wherein the PD was 23.3% and the coefficient of variation was in the range 0.2–1.0.

Similarly, the coefficient of variation for the bearing capacity safety factor against Monte Carlo iterations is depicted in Figure 5.9. The results reveal that the mean and the COV for the bearing capacity safety factor generally converge for 70 to 80 iterations or more. Therefore, the number of Monte Carlo iterations for the bearing capacity analysis was set to 100 in this study. Consequently, the variation in bearing capacity safety factor

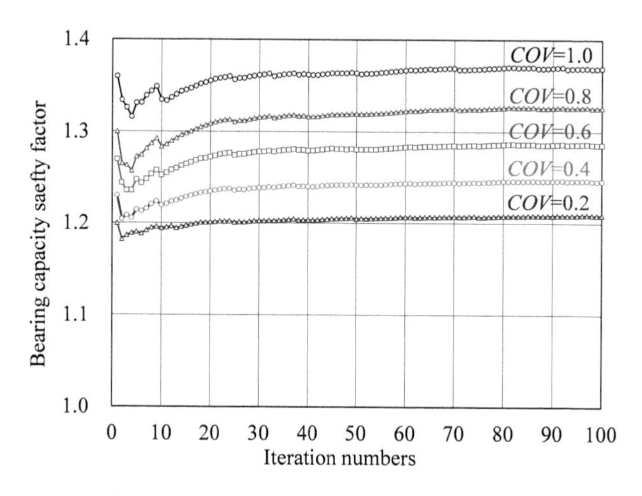

FIGURE 5.8 Mean bearing capacity safety factor against iteration numbers.

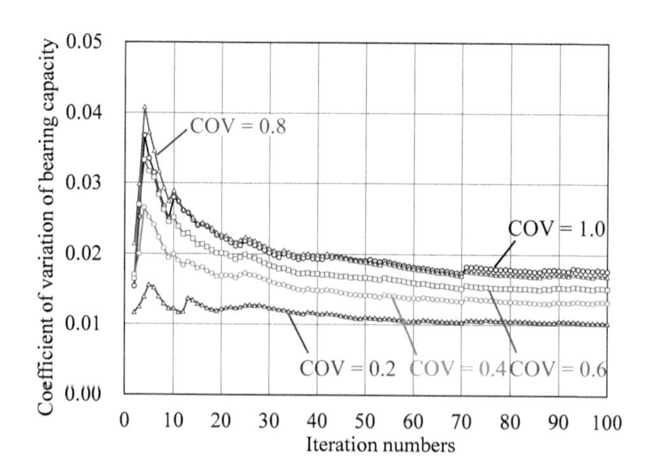

FIGURE 5.9 Coefficient of variation of bearing capacity safety factor against iteration numbers.

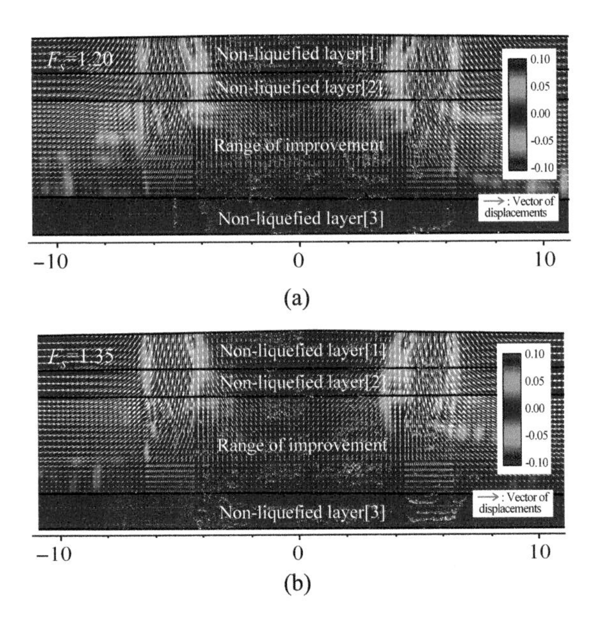

FIGURE 5.10 Maximum shear strain distribution for Cases 3a and 3e (a) Case 3a (PD: 23.3%, COV0.2), (b) Case 3e (PD: 23.3%, COV1.0).

reflecting the strength spatial variability due to chemical grouting was not substantially large, even for ten iterations. Thus, it is considered that there is no strength spatial variability in the surface non-liquefiable layer up to GL-2 m above the improvement area with varying undrained shear strength, which causes a marginal influence on the variations of the bearing capacity safety factor. This is consistent with the results reported in previous studies (Kutsuna et al., 2007).

The typical examples of maximum shear strain distributions for Case 3a and Case 3e are presented in Figure 5.10. The maximum shear strain in Case 3, with the undrained shear strength varying within the improvement area, extended deeper into the improvement area than that in Case 1 (PD = 0% and no strength variation), as shown in Figure 5.5. However, the maximum shear strain distribution between Case 3a and Case 3e did not display any significant deviation.

The typical example of deformed mesh for Cases 3a and 3e is plotted in Figure 5.11. The displacement displayed in Figure 5.11 was exaggerated by 20 times to facilitate the deformation within the improvement area. In both cases, the settlement occurred in the vicinity of the aircraft loading area, implying the extension of deformation to the improvement area. However, similar to the results of the maximum shear strain distribution,

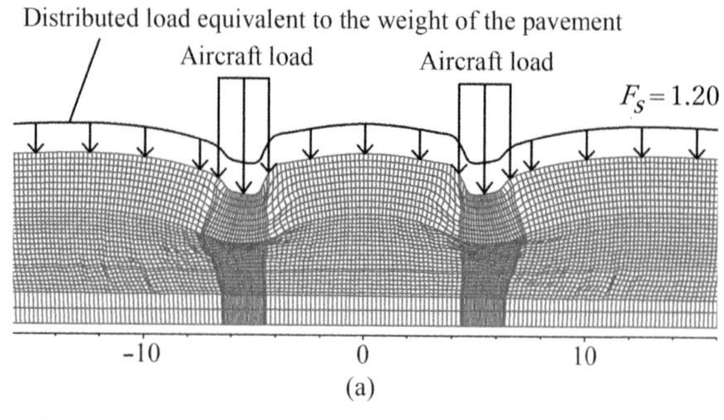

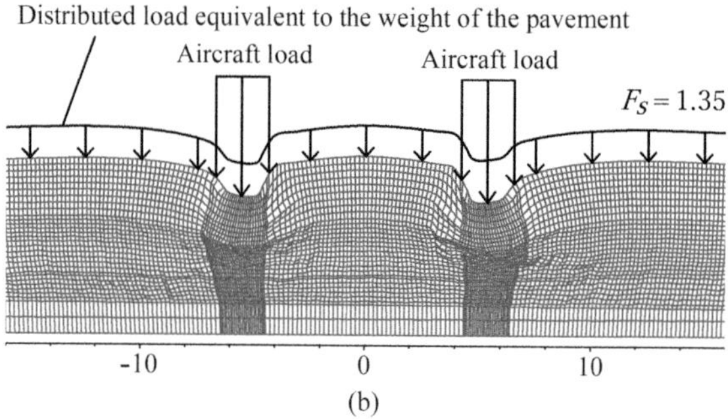

FIGURE 5.11 Deformed mesh for Cases 3a and 3e (a) Case 3a (PD: 23.3%, COV0.2), (b) Case 3e (PD: 23.3%, COV1.0).

no significant differences were observed in the deformation mode between Case 3a and Case 3e.

5.4 PERFORMANCE-BASED EVALUATION FOR BEARING CAPACITY

As discussed in Section 3, the PD range of q_u investigated for each block was 3–38%, and the COV was 0.46–0.85. Accordingly, bearing capacity analyses were conducted for nine cases of PD with COV ranging from 0.2 to 1.0, as discussed in Section 4. Based on the results of a parametric study for a given COV (=0.2, 0.4, 0.6, 0.8, and 1.0) and PD, the relationship between the bearing capacity safety factor and PD is portrayed in Figure 5.12. It is noted that the bearing capacity safety factor is set to the lower confidence limit of 99% (the reliability = 99% under the performance-based specification) to consider the safety margin in the concept of performance-based design. It can be seen that the bearing capacity safety factor decreases with increasing PD and decreasing COV. In order to satisfy the bearing capacity safety factor > 1.0, PD less than 40% is needed, irrespective of COV.

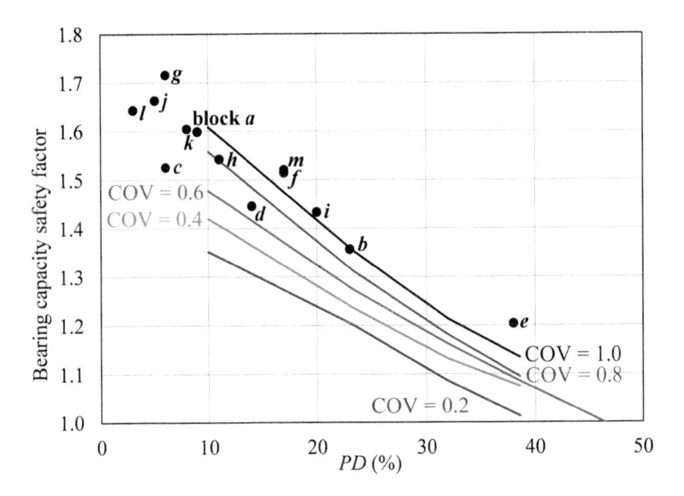

FIGURE 5.12 Bearing capacity safety factor and PD for reliability = 99%.

In addition, the black plots in Figure 5.12 were the result for the combination of PD and COV obtained by q_u site investigation for all blocks in improved ground at Airport A, as shown in Table 4.7 of Chapter 4. Note that the result for each block was obtained from another series of Monte Carlo simulations for different finite element mesh modeled by the cross-section for each block. In addition, the thickness and dimension of the improved area and original ground such as old fill, clay, and sandy layers were slightly different among all blocks, while the similar input parameters for modeling of these layers showing in Table 5.3 were used for all blocks. The bearing capacity safety factor for all blocks was greater than 1.0. Consequently, the effectiveness of ground improvement for the liquefaction countermeasure by chemical grouting at Airport A was quantitatively ensured with the reliability of 99% in terms of the bearing capacity against aircraft loading after an earthquake even if there is strength reduction of elements due to a local liquefaction resulting from strength spatial variability in chemical grouting.

Thereafter, to discuss the size effect of the ground improvement on bearing capacity together with the strength spatial variability effect, we compared the current results with those reported by Kasama et al. (2012) in which the full-replacement effect of ground improvement was evaluated for the bearing capacity under vertical loading. In addition, we accounted for the conditions stated in Kasama et al. (2019) in which a partial zone/block improvement effect beneath the footprint of the foundation was evaluated for the bearing capacity under general loading

TABLE 5.5 Comparison of Numerical Conditions with Prior Studies

	Current Study	Kasama et al. (2019)	Kasama et al. (2012)
Ground improvement	Chemical grouting	Cement mixing	Cement mixing
Improvement area	Layer improvement	Zone/block improvement	Full-replacement
Improvement width	4*B*	*B*	5*B*
Improvement depth	0.27*B*	1.25*B*	2*B*
Mean undrained shear strength of improved area	25.8–237.3 kN/m²	100, 200, 400, 800, 2000, 4000, 8000 kN/m²	100 kN/m²
COV of undrained shear strength	0.2, 0.4, 0.6, 0.8, 1.0	0.1, 0.2, 0.4, 0.8	0.2, 0.4, 0.6, 0.8, 1.0
Ratio of horizontal and vertical autocorrelation lengths	10	1.0	1.0
Autocorrection length in vertical and horizontal direction	0.2 m for vertical direction and 5.0 m for horizontal direction	Random[a], 0.15*B*, 0.5*B*, 1.0*B* for both directions	Random[a], 0.25*B*, 0.5*B*, 1.0*B*, 2.0*B*, 4.0*B* for both directions
Numerical method	Finite element analysis with shear strength reduction method	Numerical limit analyses	Numerical limit analyses
Monte Carlo iteration	100	1000	1000

Note: B: Width of foundation.

[a] Undrained shear strength for elements in finite element mesh is changed randomly.

conditions. Table 5.5 shows the comparison of numerical conditions with those of prior studies by Kasama et al. (2019) and (2012). Prior studies used the numerical limit analysis for Monte Carlo simulation with the assumption of isotropic strength autocorrelation for vertical and horizontal directions. Current study uses SSR-FRM for Monte Carlo simulations with 0.2 m and 5.0 m autocorrelation length for vertical and horizontal directions, respectively.

The mean undrained shear strength of actual ground improved by chemical grouting for current study ranges from 25.8 k to 237.3 kN/m², which is equivalent to a low strength range compared to that of zone/block improvement by cement mixing used in the parametric study by Kasama et al. (2019), 100–8000 kN/m². The range of mean undrained shear strength of improved ground by full replacement of cement mixing is assumed to be 100 kN/m^2 in the parametric study of Kasama et al. (2019). The range of COV for undrained shear strength is covered from 0.1 to 1.0 for three studies.

Although the input parameters for three studies are slightly different as described above, the size effect of improvement area on the bearing capacity reduction due to strength variability was carefully investigated by using the results of prior studies that match the input parameter condition to that of current study as much as possible. Namely, the numerical result for the mean undrained shear strength of 100 kN/m^2 and autocorrelation length = Random (smallest autocorrelation length case; the strength of finite element mesh was randomly determined) in Kasama et al. (2012) was utilized, while that for the mean undrained shear strength of 100–200 kN/m^2 and autocorrelation length divided by foundation width $\theta/B = 0.15$ (smallest autocorrelation length case) in Kasama et al. (2019) was used for comparison.

The improvement area of this study with those of the two previous studies, Kasama et al. (2012) and Kasama et al. (2019), is presented in Figure 5.13; the influential area of undrained shear strength spatial variability on the bearing capacity of the shallow foundation proposed by Otake and Honjo (2012) (hereinafter, "the range to be locally averaged (*RLA*)") is also presented. *RLA* is ~2 times the width of the foundation *B* in the horizontal direction and ~0.7 times in the vertical direction. Notably, the improvement size ratio *ISR* is defined as the size ratio of improvement area to *RLA*.

In Kasama et al. (2012), the improvement width and depth were assumed to be 5 and 2 times the foundation width *B*, respectively, indicating an improvement of 100% of the *RLA* (*ISR* = 1.0). In Kasama et al. (2019), an

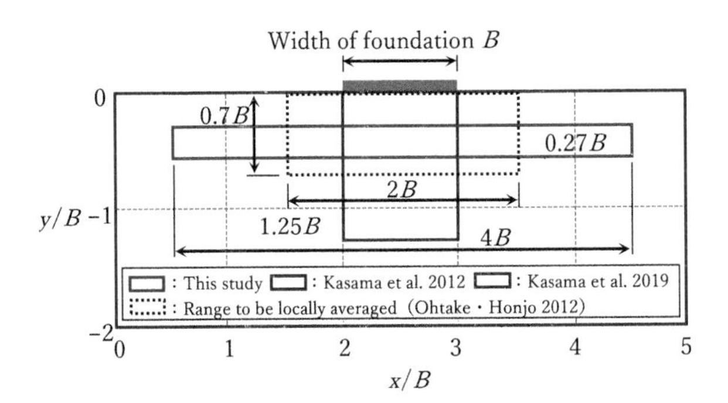

FIGURE 5.13 Comparison of improvement areas among this study and previous studies.

improvement width was equivalent to the foundation width, while an improvement depth was 1.25 times the foundation width, which implied an improvement of ~50% of the *RLA* ($ISR = 0.5$). In this study, the improvement depth was approximately 4 m and the aircraft load width was assumed as ~15 m for a runway width of 60 m. Therefore, this 15 m width is assumed as the foundation width. As a result, 39% of the *RLA* ($ISR = 0.39$) was improved, which is the lowest improvement area in the *RLA* compared to the two previous studies. It is noted that the results of Kasama et al. (2019) for correlation length divided by foundation width = 0.15 and Kasama e al. (2012) for correlation length = Random are used for comparison.

In the general design of a surface strip foundation with a width, B, on improved ground, the ultimate bearing capacity $Q_{\text{hom}} = B \times \mu_c \times N_{\text{chom}}$, in which the mean undrained shear strength μ_c and the bearing capacity factor N_{chom} are based on the simplified failure mode obtained for a homogeneous improved ground. In order to consider the safety margin in design, the ultimate bearing capacity Q_{hom} is decreased (using the safety factor, FS), such that the allowable bearing capacity, $Q_a = B \times \mu_c \times N_{\text{chom}}/\text{FS}$. It is emphasized that N_{chom} is assumed to be a constant value and FS is considered to incorporate the variabilities of soil property and the uncertainty of external force to the foundation as well as the model error in design.

On the other hand, the ultimate bearing capacity Q for spatially varied ground can then be found from $Q - B \times \mu_c \times N_c$ when the bearing capacity factor N_c for improved ground is a probabilistic parameter. From the current calculations, spatial variability sets to reduce the bearing capacity of improved ground and $Q = B \times \mu_c \times N_c$ can be less than $Q_{\text{hom}} = B \times \mu_c \times N_{\text{chom}}$ for the homogeneous improved ground. Even though Q is expected to be larger than the allowable bearing capacity $Q_a = B \times \mu_c \times N_{\text{chom}}/\text{FS}$, which is reduced by dividing Q_{hom} with FS, there is a probability showing that $Q = B \times \mu_c \times N_c$ is less than $Q_a = B \times \mu_c \times N_{\text{chom}}/\text{FS}$. In order to discuss appropriate FS to ensure that $Q > Q_a$, the probability $P[Q < Q_a]$ is given by the following equation, assuming that N_c of cement-treated ground is described by a log-normal distribution:

$$P[Q < Q_a] = P[Q < Q_{\text{hom}}/FS] = \Phi\left(\frac{\ln\left(N_{\text{chom}}/FS\right) - \mu_{\ln N_c}}{\upsilon_{\ln N_c}}\right) \quad (5.8)$$

where $\Phi(..)$ is the cumulative normal function and FS is a safety factor. N_{chom} is the bearing capacity computed for the case where the undrained shear strength of the improved zone is homogeneous. The $\mu_{\ln Nc}$ and $\sigma_{\ln Nc}$ are the mean and standard deviation of ln N_c obtained by the following equations using $COV_{Nc} = \sigma_{Nc}/\mu_{Nc}$.

$$\sigma_{\ln N_c} = \sqrt{\ln\left(1 + COV_{N_c}^2\right)} \tag{5.9a}$$

$$\mu_{\ln N_c} = \ln \mu_{N_c} - \frac{1}{2}\sigma_{\ln N_c}^2 \tag{5.9b}$$

Notably, the uncertainty of the external force on the foundation is not considered in Equation (5.8). In order to compare $P[Q < Q_a]$ with those for the current design code, the estimated probabilities of failure considered in load and resistance factor design (LRFD) codes for shallow foundations are reported in the range, $P_f = 10^{-2} - 10^{-3}$ (Baecher and Christian, 2003; Phoon et al., 2000). The FS value is discussed to satisfy $P[Q < Q_a]$ with $P_f = 10^{-2}–10^{-3}$ in the following section.

It is important to connect the conventional load and resistance factor design to reliability-based and performance-based design and understand their connection. Namely, from the viewpoint of the bearing capacity reduction due to the spatial variability and improvement size, the factor of safety in the conventional load and resistance factor design was discussed to satisfy the probability of failure $P_f = 10^{-2} - 10^{-3}$. Therefore, in order to investigate the relationships between $P[Q < Q_a]$ and FS for the bearing capacity under vertical loading conditions, Figure 5.14 summarizes $P[Q < Q_a]$ against

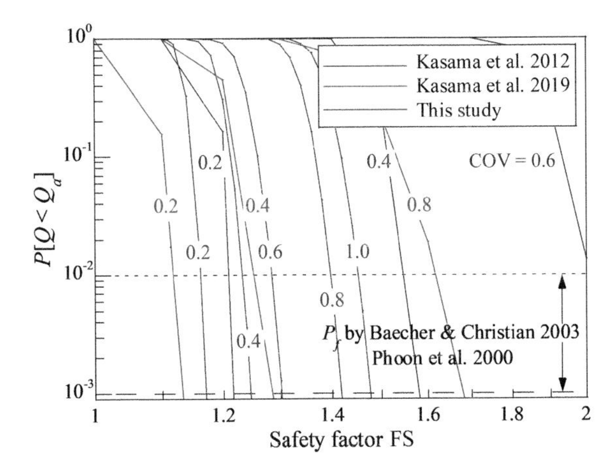

FIGURE 5.14 Relationship between safety factor and failure probability.

FS for different improvement areas (different ISR) as a function of COV. It can be seen that $P[Q < Q_a]$ decreases markedly with increasing FS irrespective of COV, but also intensifies with increasing COV for a given ISR.

Therefore, in order to satisfy the required probability of failure such as $P_f = 10^{-2}$–10^{-3}, a safety factor for a given ISR should be properly determined depending on COV. In addition, $P[Q < Q_a]$ also depends on values of ISR, namely, the probability for a given safety factor increases as ISR increases in the current analysis. Since the conventional safety factor FS for shallow foundation is 2.0–3.0, the results in Figure 5.14 show that $P[Q < Q_a]$ for $ISR = 0.39$ and 0.5 is much less than $P_f = 10^{-2}$–10^{-3} for COV = 0.2 – 1.0, while $P[Q < Q_a]$ for $ISR = 1.0$ is beyond $P_f = 10^{-2} - 10^{-3}$ for COV > 0.6. It can be characterized that the size of the improvement area has a large impact on the probability $P[Q < Q_a]$ for bearing capacity, meaning the probability $P[Q < Q_a]$ decreases with decreasing ISR and COV. Namely, a small improvement area with strength spatial variability provides a small failure probability for bearing capacity.

Figure 5.15 illustrates the safety factor, FS, for designs of improved ground (with specified parameters, COV and ISR for vertical loading and small correlation length) in order to satisfy $P[Q < Q_a]$ with $P_f = 10^{-2}$ and 10^{-3}. It can be seen that the largest values of FS occur for $ISR = 1.0$ and that FS increases with COV. It can be suggested that FS should change according to the spatial variability and improvement area size of improved ground. Typical improved ground with a large improvement area and a large degree of variability (ISR = 1.0, COV ≥ 0.6) will require FS > 2.0 to satisfy $P_f = 10^{-2}$, while FS = 1.5 is appropriate for improved ground with a small improvement area and small spatial variability (ISR < 0.5 and COV = 0.2–0.4 similar to naturally deposited soils). Specifically, for small

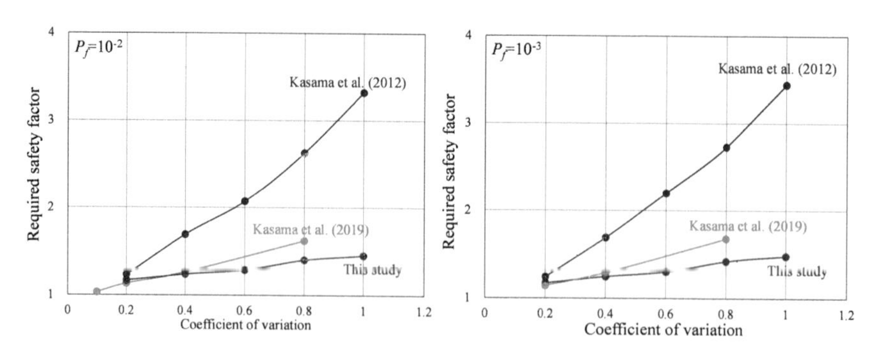

FIGURE 5.15 Relationship between required safety factor and coefficient of variation.

improvement area (ISR = 0.39 in current study), the safety factor dependency on COV is very small and the safety factor ranges from 1.1 to 1.3. Compared to the results of the two studies, Kasama et al. (2012) (*ISR* = 1.0) reported the largest factor of safety FS required, and the required FS diminished in the order of Kasama et al. (2019) for vertical zone/block-type improvement and the current study for a layered-type improvement. As the improvement size ratio *ISR* decreases, the safety factor FS required for the target-bearing capacity decreases.

From the viewpoint of quality control and assurance for improved ground, the concept of percentage of defective (PD) is used instead of safety factor FS. The definition of PD is the percentage of defective samples having a quality level below the specific quality. For example, the concept of percentage of defective has been used for the quality control of ready-mixed concrete, and PD less than 5% is conventionally required in the quality control and assurance of concrete strength, as reported in JSCE (2007) and ACI (2011). Figure 5.16 shows the PD for improved ground with a small improvement area (*ISR* = 0.39 in the current study) needed to achieve $P_f = 10^{-2}$ and 10^{-3} as a function of COV. It can be seen that COV has much more influence on PD. Namely, PD expands with increasing COV. Specifically, the minimum PD for both $P_f = 10^{-2}$ and 10^{-3} occurs at COV = 0.2 with PD ≈ 23%. Therefore, it is emphasized that the allowable PD for a layered-type improved ground by chemical grouting is suggested to be 23% for $P_f = 10^{-2}$ and 10^{-3}. In prior studies from Kasama et al., 2012 and 2019, the allowable PD values for full-replacement cement-treated ground were suggested as 10% and 5% for $P_f = 10^{-2}$ and 10^{-3}, respectively, and those for block-type cement-treated ground were suggested to be 15% and 10% for $P_f = 10^{-2}$ and 10^{-3}, respectively. Namely, the PD values for layered-type improved ground with small improvement size are 13–18%

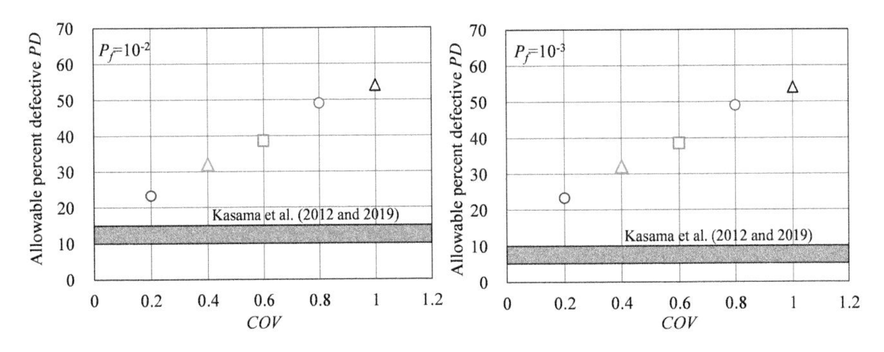

FIGURE 5.16 Allowable percent defective against COV.

and 8–13% larger than those reported for full-replacement improved ground and block-type improvement ground, respectively.

5.5 SUMMARY

In this chapter, we conducted a bearing capacity analysis considering strength spatial variability and employed a statistical method to evaluate the bearing capacity for the ground improved by the chemical grouting based on the performance specifications. We arrived at the following conclusions:

1) From the results of parametric study of bearing capacity analysis for layered-type improved ground by the chemical grouting with variability in undrained shear strength, allowable percentage of defective samples from the viewpoint of practical quality control and assurance should generally be less than 40% so that the bearing capacity safety factor exceeds 1.0.
2) The effectiveness of the chemical grouting as the liquefaction countermeasure at Airport A was quantitatively ensured with the reliability of 99% in terms of the bearing capacity against aircraft loading after an earthquake, even if there is strength reduction of elements due to local liquefaction resulting from strength spatial variability in chemical grouting.
3) A higher strength variability (coefficient of variation) yields the requirement for a higher safety factor with respect to the target-bearing capacity; however, this tendency is independent of the improvement size under the chemical grouting. More importantly, a smaller improvement size subject to strength variability requires a lower safety factor for the bearing capacity and higher allowable percentage defective. Consequently, the allowable percentage of defective samples for a layered-type improved ground by chemical grouting is suggested to be 23% for both $P_f = 10^{-2}$ and 10^{-3}, respectively.

REFERENCES

American Concrete Institute, 2011. Guide to Evaluation of Strength Test Results of Concrete (ACI 214R-11) Reported by ACI Committee 214.

Baecher, G.B., Christian, J.T., 2003. *Reliability and Statistics in Geotechnical Engineering*. John Wiley & Sons, New York.

Chen, E.J., Liu, Y., Lee, F.-H., 2016. A statistical model for the unconfined compressive strength of deep-mixed columns. *Géotechnique* 66 (5), 351–365. https://doi.org/10.1680/jgeot.14.P.162

Civil Aviation Bureau, Ministry of Land, Infrastructure, Transport and Tourism, 2019a. *Design Guidelines and Design Examples for Airport Civil Engineering Facilities (Seismic Design Version).*

Civil Aviation Bureau, Ministry of Land, Infrastructure, Transport and Tourism, 2019b. *Design Guidelines and Design Examples for Airport Civil Engineering Facilities (Structural Design).*

Griffiths, D.V., Fenton, G.A., 2001. Bearing capacity of spatially random soil: The undrained clay Prandtl problem revisited. *Géotechnique* 51, 351–359.

Griffiths, D.V., Fenton, G.A., Manoharan, N., 2002. Bearing capacity of rough rigid strip footing on cohesive soil: Probabilistic study. *J. Geotech. Geoenviron. Eng.* 128, 743–755.

Honjo, Y., 1982. A probabilistic approach to evaluate shear strength of heterogeneous stabilized ground by deep mixing method. *Soils Found.* 22(1), 23–38.

Japanese Society of Civil Engineers, 2007. *Guidelines for Concrete No.16, Standard Specifications for Concrete Structures 2007, Materials and Construction.* http://www.jsce.or.jp/committee/concrete/e/JGC16_Standard%20Specifications_Materials%20and%20Construction_1.1.pdf

Jellali, B., M. Bouassida, de Buhan, P., 2005. A homogenization method for estimating the bearing capacity of soils reinforced by columns. *Int. J. Numer. Anal. Methods Geomech.* 29 (10), 989–1004. https://doi.org/10.1002/nag.441

Kasama, K., Whittle, A.J., 2011. Bearing capacity of spatially random cohesive soil using numerical limit analysis. *J. Geotech. Geoenviron. Eng. ASCE* 137 (11), 989–996.

Kasama, K., Whittle, A.J., Kitazume, M., 2019. Effect of spatial variability of block-type cement-treated ground on the bearing capacity of foundation under inclined load. *Soils Found.* 59, 2125–2143.

Kasama, K., Whittle, A.J., Zen, K., 2012. Effect of spatial variability on the bearing capacity of cement-treated ground. *Soils Found.* 52, 600–619.

Kasama, K., Zen, K. and Whittle, A.J., 2006. Effects of spatial variability of cement-treated soil on undrained bearing capacity, *Proceedings of International Conference on Numerical Modeling of Construction Processes in Geotechnical Engineering for Urban Environment*, Bochum, Germany, 305–313.

Kasama, K., Zen, K. and Whittle, A.J., 2008. Bearing capacity of clayey soil using random field numerical limit analyses, *J. Applied Mechanics JSCE*, 291–298.

Kobayashi, M., 1984. Stable Analysis of Geotechnical Structures by Finite Method, Report of the Port and Harbour Research Institute, 23, 83–101.

Kutsuna, A., Zen, K., Chen, G., Kasama, K., 2007. Numerical limit analysis on the bearing capacity considering the locality of liquefaction and seismic loading. *Proc. JSCE Earthquake Eng. Symp.* 29, 331–335.

Larsson, S., Stille, H., Olsson, L., 2005. On horizontal variability in lime-cement columns in deep mixing. *Géotechnique* 55 (1), 33–44. https://doi.org/10.1680/geot.2005.55.1.33

Liu, Y., He, L.Q., Jiang, Y.J., Sun, M.M., Chen, E.J., Lee, F.H., 2019. Effect of in situ water content variation on the spatial variation of strength of deep cement-mixed clay. *Géotechnique* 69 (5), 391–405. https://doi.org/10.1680/jgeot.17.P.149

Liu, Y., Jiang, Y.J., Xiao, H., Lee, F.H., 2017. Determination of representative strength of deep cement-mixed clay from core strength data. *Géotechnique* 67 (4), 350–364. https://doi.org/10.1680/jgeot.16

Lyamin, A.V., Sloan, S.W., 2002. Lower bound limit analysis using non-linear programming. *Int. J. Numer. Methods Eng.* 55, 573–611.

Matthies, H.G., Brenner, C.E., Bucher, C.G., Guedes Soares, C., 1997. Uncertainties in probabilistic numerical analysis of structures and solids – Stochastic finite elements. *Struct. Saf.* 19, 283–336.

Namikawa, T., 2016. Conditional probabilistic analysis of cement-treated soil column strength. *Int. J. Geomech.* 16 (1), 04015021. https://doi.org/10.1061/(ASCE)GM.1943-5622.0000481

Namikawa, T., Junichi Koseki, J., 2013. Effects of spatial correlation on the compression behavior of a cement-treated column. *J. Geotech. Geoenviron.*, 139 (8), 1346–1359. https://doi.org/10.1061/(ASCE)GT.1943-5606.0000850

Navin, M.P., Filz, G.M., 2005. Statistical analysis of strength data from ground improved with DMM columns. In: *Proceedings International Conference on Deep Mixing Best Practice and Recent Advances (Deep Mixing'05), CD-ROM*, Stockholm, Sweden.

Otake, Y., Honjo, Y., 2012. A practical geotechnical reliability based design employing response surface -seismic design of irrigation channel on liquefiable ground. *J. JSCE Ser. C.* 68, 68–83.

Pan, Y., Liu, Y., Xiao, H., Lee, F.H., Phoon, K.K., 2018. Effect of spatial variability on short- and long-term behaviour of axially-loaded cement-admixed marine clay column. *Comput. Geotech.*, 94, 150–168. https://doi.org/10.1016/j.compgeo.2017.09.006

Phoon, K.K., Kulhawy, F.H., Grigoriu, M.D., 2000. Reliability-based design for transmission line structure foundations. *Comput. Geotech.* 26, 169–185.

Popescu, R., Deodatis, G., Nobahar, A., 2005. Effects of random heterogeneity of soil properties on bearing capacity. *Probab. Eng. Mech.* 20, 324–341.

Sloan, S.W., Kleeman, P.W., 1995. Upper bound limit analysis using discontinuous velocity fields. *Comput. Methods Appl. Mech. Eng.* 127, 293–314.

Terashi, M., Tanaka, H. 1981. Ground improved by deep mixing method. *Proceedings of. 10th ICSMFE, Stockholm*, 3, 777–780.

The Coastal Development Institute of Technology, 2018. Coastal Technology Library No. 29 Technical Manual for Deep Mixing Treatment Method in Offshore Construction.

Vanmarcke, E.H., 1984. *Random Fields: Analysis and Synthesis.* MIT Press, Cambridge, Massachusetts.

Zen, K., Yamazaki, H., Yoshizawa, H. and Mori, K., 1992. Development of premixing method against liquefaction, *Proceedings of 9th Asian Regional Conference, SMFE*, 1, 461–464.

Zienkiewicz, O.C., Humpheson, C., Lewis, R.W., 1975. Associated and nonassociated visco-plasticity and plasticity in soil mechanics. *Géotechnique* 25, 671–689.

CHAPTER 6

Performance-Based Verification on Earthquake-Induced Deformation

Tomoyuki Kaneko

6.1 INTRODUCTION

Liquefaction of the ground is one of the causes of damage to civil engineering structures such as ports, airports, river embankments, and road embankments in the event of large-scale earthquakes. Ground improvement is necessary as a countermeasure against liquefaction. Ground improvement methods can be roughly classified into three categories based on the principle of improvement: density-increasing methods, consolidation-accelerating methods, and consolidation methods. Among these methods, the consolidation method causes large variations in soil constants such as elastic modulus and shear strength constants due to heterogeneity in soil properties and mixing and compaction of consolidation materials (Zen et al., 1990; Kasama et al., 2010; Hatanaka and Murono, 2002; Suetomi et al., 2000). This spatial heterogeneity in soil properties is expected to cause local liquefaction and shear failure during earthquakes, which may affect bearing capacity and deformation modes.

In an experimental study on the dynamic deformation of improved soil, Zen et al. (1990) found that in cyclic triaxial tests on sandy soil

DOI: 10.1201/9781032670133-6

improved with cement and other solidifiers, the specimens failed in shear when the strength of the specimens was high and failed similarly to liquefaction when the strength was low. The authors (Kasama et al., 2010) also conducted shaking table tests in a gravity field that reproduced the heterogeneity of the improved soil in a simplified manner and discussed the additional effect of reducing post-seismic settlement beyond the presence of improvement elements using an index called liquefaction clusters introduced based on percolation theory.

In a study of heterogeneity of the Permeation Grouting Method (PGM) using numerical analysis and probabilistic statistical methods, Hatanaka and Murono (2002) conducted Monte Carlo simulations using a probabilistic soil model set up with a one-dimensional finite element method to investigate the effect of spatial heterogeneity of soil properties on the behavior of soil during earthquakes and clarified the effect of heterogeneity in shear wave velocity and internal friction angle on earthquake response. As studies of heterogeneous soils using a two-dimensional finite element method, Suetomi et al. (2000), studied the scattering and attenuation characteristics of soil surface layers modeled by the spatial distribution of heterogeneous soil properties, and Nakamura et al. (2007) considered the effects on the variation of the strong nonlinear response of soils due to differences in the modeling of the spatial distribution of heterogeneous soil properties. Tsutida and Ono (1988) proposed a method to estimate the amount of unequal settlement due to compaction of clay soil at airport facilities by a numerical simulation program that evaluates the variation of ground constants and layer thickness in a three-dimensional space.

Miyata et al. (1998) proposed a method for estimating unequal settlement due to liquefaction based on a probabilistic evaluation of the spatial variation of geotechnical properties at airport facilities. Honjo and Otake (2012) and Otake and Honjo (2012) proposed a method to evaluate the variation of geotechnical parameters and statistical estimation errors in reliability analysis of geotechnical structures and developed a practical reliability design method for geotechnical structure design using the evaluation of a channel on liquefied ground as an example. Kataoka et al. (2011) proposed a liquefaction risk analysis method for PGM based on liquefaction probabilities obtained from Monte Carlo simulations using nonlinear seismic response analysis considering heterogeneity of geotechnical properties of the PGM, regional characteristics and uncertainties of external earthquake forces, and economic losses caused by local liquefaction.

Juang et al. (2005) considered the liquefaction probability of soil based on the results of standard penetration tests conducted on the original soil and proposed an equation relating the cyclic shear strength ratio to the liquefaction potential. Popescu et al. (2005) used a three-dimensional finite element method to represent geotechnical heterogeneity and simulated localized liquefaction caused by such heterogeneity. The influence of spatial heterogeneity of soil properties on ground behavior during earthquakes is becoming clearer.

This chapter focuses on the ground improved by PGM, which is one of the liquefaction countermeasure methods for solidification systems. As with other PGM, the infiltration heterogeneity of the chemical solution and the heterogeneity of the properties in the soil are the main reasons for the spatial heterogeneity of the material constants of the infiltrated soil. The heterogeneity of the PGM is expected to affect the bearing capacity and deformation of the PGM at the time of design, and an assessment for bearing capacity and deformation based on performance specifications that takes the heterogeneity of the soil into account is required in practical terms. However, no previous studies have examined the heterogeneity of permeation-treated soils in detail using actual data, nor are there any that proposed bearing capacity and deformation evaluation methods based on performance specifications that assume variability. The authors (Kasama et al., 2022) conducted a detailed verification of the heterogeneity of shear strength of improved ground using actual data on improved ground for the purpose of liquefaction countermeasures at airports. The bearing capacity analysis was performed by Monte Carlo simulation using the finite element method and the shear strength reduction method.

In this chapter, the same improved soil is used to analyze the deformation behavior of the soil due to dissipation of excess pore water pressure after liquefaction by expressing the shear strength in random field theory and modeling the liquefaction of the soil under earthquake conditions in a simplified manner. The results of the analysis are used to examine probabilistically and statistically the effect of heterogeneity of the improved soil on the deformation of airport runways after earthquakes, and a new performance evaluation method based on performance specifications is proposed for the bearing capacity and flatness required of airport runways after earthquakes. The scope of the soil improved by chemical grouting, the results of the post-earthquake investigations, and the setting conditions such as the location of the analytical cross-section are the same as in the previous study (Kasama et al., 2022). It should also be noted that the heterogeneity was considered in the analysis only for the unconfined compression strength of the area to be improved in the PGM, and the variation in the surrounding unimproved soil and target earthquake ground motion was not taken into account, as in the previous study (Kasama et al., 2022).

6.2 DEFORMATION ANALYSIS OF GROUND IMPROVED BY CHEMICAL GROUTING

6.2.1 Overview of Deformation Analysis

When verifying the gradient of basic airport facilities such as runways, it is necessary to determine the deformation of the foundation ground of the facility during an earthquake. Seismic forces acting on the facility and liquefaction of the foundation ground are two factors that may cause damage to airport facilities during an earthquake. According to the *Airport Seismic Design Guidelines* (MLIT, 2019), past earthquake damage cases of airport facilities have confirmed that gradient deviations and steps rarely occur if the foundation ground is not liquefied. In other words, the main cause of seismic deformation of basic airport facilities is the reduction of ground stiffness due to liquefaction and the dissipation of excess pore water pressure. Therefore, it is necessary to use an analytical method that can predict the residual deformation after an earthquake—which is mainly caused by liquefaction of the ground—to verify the gradient of basic airport facilities.

Typical analysis methods that can predict residual deformation mainly due to liquefaction can be roughly classified into the dynamic analysis method (Iai et al., 1992; Oka et al., 1994), which tracks the ever-changing seismic deformation by seismic response analysis, and the simple static analysis method (Yasuda et al., 1999; Uzuoka et al., 1997), which sets appropriate post-earthquake ground properties (stiffness reduction). In this chapter, since the Monte Carlo simulation considering the heterogeneity of the improved ground involves several hundred cases, the seismic response analysis usually performed in the seismic performance verification is not performed, but a simplified static analysis is used for the verification in consideration of the computational load. Since it is assumed no aircraft are traveling during the earthquake and the modulus of elasticity is not significantly different before and after the chemical grouting (CDIT, 2020), it is assumed that complex seismic response does not occur in the heterogeneous ground after the improvement, and a two-dimensional seismic response analysis involving several hundred cases for the design earthquake motion is not conducted. Instead, deformation verification was performed using static analysis, which requires a relatively short analysis time. In other words, the distribution of shear stress due to seismic propagation during the earthquake and the process of excess pore pressure increase during the earthquake are not considered.

Table 6.1 shows the analytical modeling methods used in this chapter for the following phases: (a) before the earthquake, (b) during the earthquake, and (c) immediately after the earthquake to several days after the earthquake. The results of field and laboratory permeability tests at Airport A—the subject of this study—showed that the hydraulic conductivity of the liquefied layer—the layer to be improved—was about 9.5×10^{-3} cm/sec, and that of the improved layer was about 1.2×10^{-6} cm/sec. Based on the results of the excess pore water pressure dissipation analysis described below, it was assumed that the excess pore water pressure would remain even after the runway was put into service three days after the earthquake. In a previous study (Kasama et al., 2022), it was confirmed that the bearing capacity was secured under the condition of residual excess pore water pressure. In this chapter, the analysis is focused on the flatness of the runway, especially during the process of dissipation of excess pore water pressure. The specific analytical procedure is shown below.

1) A one-dimensional seismic response analysis (SHAKE) (Shnabel et al., 1972) using the equivalent linearization method with the design earthquake motion as the input condition is separately conducted to determine whether liquefaction occurs based on the external forces generated in the ground, and the necessary design base strength (unconfined compression strength) is set for the range where liquefaction occurs.

2) The average unconfined compression strength and coefficient of variation of the post-improved ground obtained from the results of the ground investigation after the ground improvement works are obtained.

3) The liquefaction was assumed not to occur at the points above the design base strength. On the other hand, at locations where liquefaction occurs below the design strength, full liquefaction is assumed to occur, and excess pore water pressure associated with liquefaction is assumed to accumulate, resulting in deformation after the earthquake due to dissipation of the excess pore water pressure.

TABLE 6.1 Each Scenario and Analytical Modeling Method

Scenario	(a) Before the Earthquake	(b) During the Earthquake	(c) After the Earthquake - A Few Days Later
	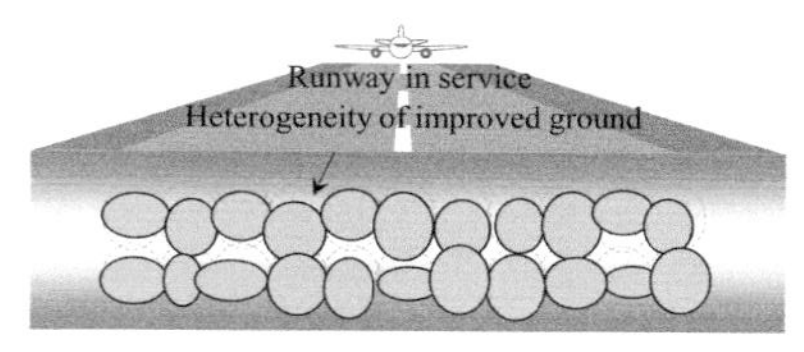	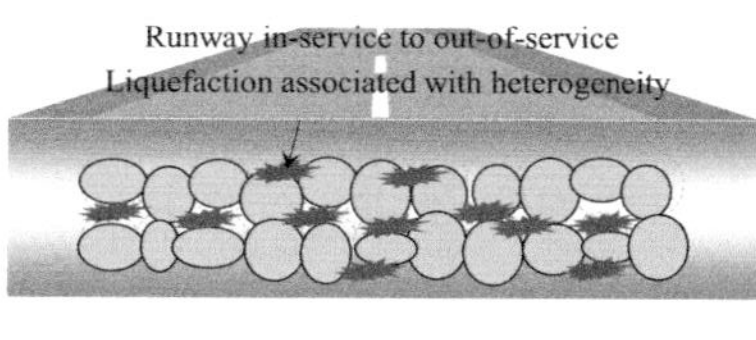	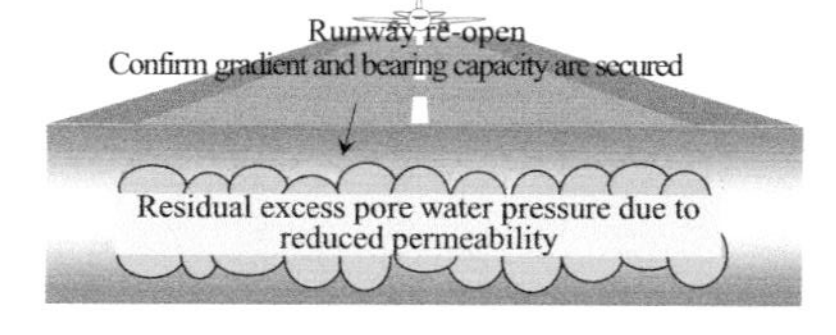
Analysis modeling method	• The modeling of the areas where the strength is partially less than the specified design strength (Black elements in the figure (d)) based on the random field theory. • 100 cases of geotechnical models with different variations of unconfined compression strength. • Strength heterogeneity is not considered for unimproved ground outside the runway area. 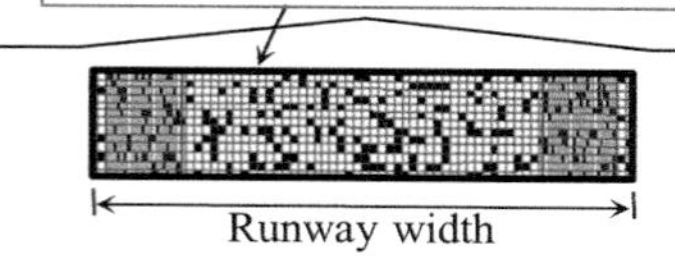	• Variations in seismic forces are not considered. • Specified design strength based on one-dimensional seismic response analysis for L2 earthquake motion. • The excess pore water pressure ratio is 0.0 for elements greater than the specified design strength, and 1.0 for elements less than the specified design strength. 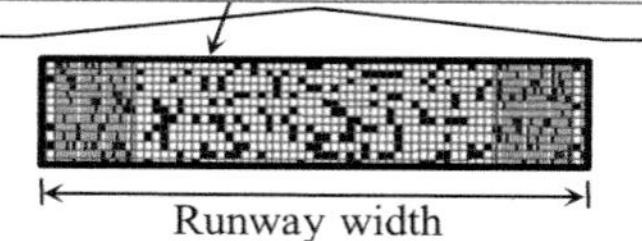	• Runway deformation (gradient) is verified by analysis of excess porewater pressure dissipation with aircraft loads immediately after the earthquake. • Assumed maximum load including the impact of aircraft running (load variation is not considered). 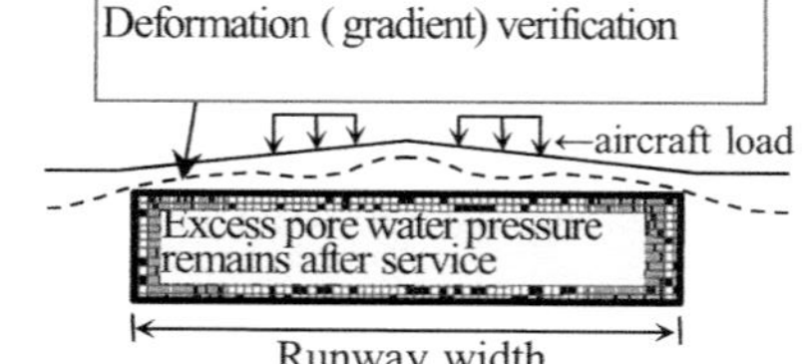

For the evaluation of the shear strength variation in the improvement range, the random field theory was used to represent spatial heterogeneity. "Airports important for air transportation" are required to be able to operate scheduled commercial flights within three days after a Level 2 earthquake. In this chapter, in order to verify the performance of this method, deformation analysis was conducted for each compliance ratio of the ground improved by the chemical grouting to obtain the runway gradient, assuming that the aircraft load is applied during the dissipation process of excess pore water pressure generated by liquefaction, as shown in Figure 6.1. In the deformation verification of this chapter, an FEM program incorporating the exponential model of "FLIPDIS (Morio et al., 2011)" was used. The program solves the dissipation process of excess pore water pressure by Biot's multidimensional consolidation equation (Biot, 1941) and calculates the soil deformation based on the effective stress coupled to the pore water pressure.

In order to represent soil softening associated with excess pore water pressure, a model is employed in which the deformation coefficient is reduced based on the following Equation (6.1), and the stress-strain relationship of the soil is nonlinear.

$$E = E_0\left(\sigma_{m0}'/\sigma_{ma}'\right)^{1/2}/\left(\exp\left(aR-b\right)+1.0\right) \tag{6.1}$$

where,

E: Modulus of elasticity at each calculation step

E_0: Modulus of elasticity under reference mean effective confining pressure

σ_{m0}': Initial average effective confining pressure

σ_{ma}': Reference average effective confining pressure

R: Excess pore water pressure ratio

a, b: experimental constant.

The reproducibility of this constitutive equation for predicting deformation after liquefaction was confirmed in an experimental study (Sugano and Nakazawa, 2009) conducted at Ishikari Bay New Port using full-scale airport facilities. In this study, a two-dimensional seismic response analysis based on the results of laboratory tests on field samples was conducted under the assumption that the airport had been homogeneously improved by chemical grouting, and a dissipation analysis of excess pore

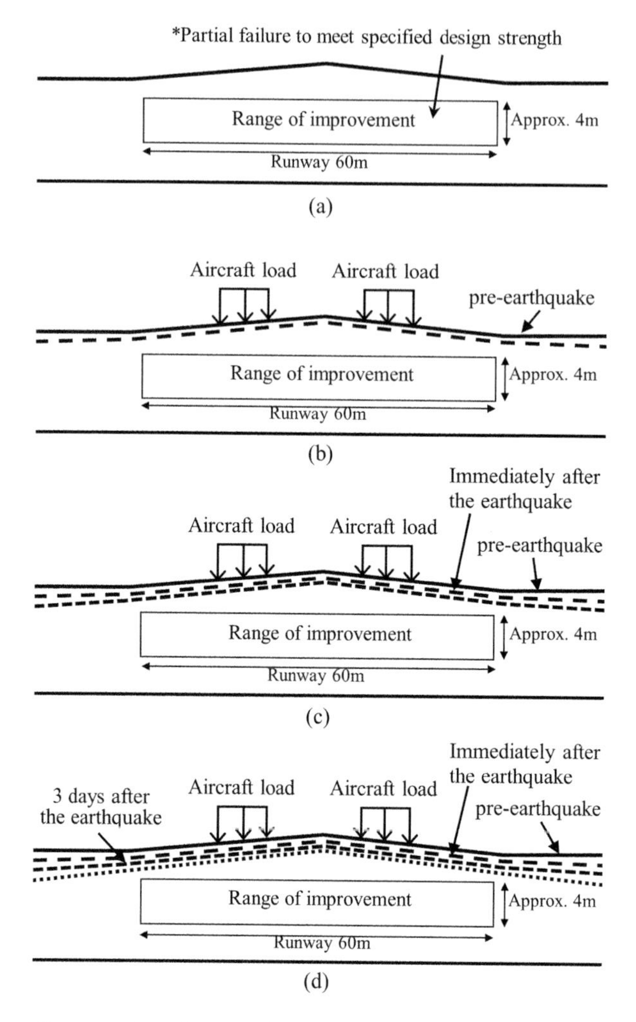

FIGURE 6.1 Image of verification by deformation analysis (a) before to during earthquake, (b) immediately after the earthquake, (c) three days after the earthquake, (d) after dissipation of excess pore water pressure.

water pressure was conducted using this constitutive equation. In this study, however, it should be noted that heterogeneous improved ground is assumed and that a two-dimensional seismic response analysis involving several hundred cases is not performed, but a static analysis is used for deformation verification, which requires a relatively short analysis time.

First, an initial stress analysis was conducted for the pre-earthquake condition without excess pore water pressure, and the effective stress of each element was analyzed. In a subsequent step, the excess pore water pressure of elements that reach an excess pore water pressure ratio of 1.0 due to full liquefaction is increased to a value equivalent to σ_v' to represent the soil immediately after liquefaction.

6.2.2 Conditions for Deformation Analysis

A typical cross-section of improved ground by chemical grouting at Airport A was modeled for deformation analysis. The airport is located on a plain with hills and terraces and a river flowing between them. In the area under study, the stratigraphic structure consists of a 2 m layer of old fill B, a 1 m layer of low permeability clay, and a 4 m layer of loose sandy soil. The fine-grain content F_c of the soil to be improved was approximately 5 to 25%, the porosity n was 39 to 44%, the N value was 3 to 15, and the liquefaction strength R_{120} was 0.27 to 0.32. The liquefaction strength of the soil to be improved was 0.27 to 0.32. The design earthquake ground motion was assumed to be the Level 2 earthquake motion set at Airport A. The maximum shear stress ratio based on one-dimensional seismic response analysis for the liquefied area was about 0.35, and the design base strength (unconfined compression strength) required for the improved ground by chemical grouting was 60 kN/m^2. The improvement area covered the entire 60 m width of the runway, and the improvement depths ranged from GL-2 m to GL-8 m.

Based on the results of the post-improvement ground investigation, a deformation analysis was conducted for the unconfined compression strength q_u within the improvement area by the chemical grouting method using the random field theory to express the variation in the assumed range of compliances and coefficients of variation under aircraft loading conditions. As shown in Figure 6.2, the unconfined compression strength q_u of the ground improved by chemical grouting was assumed to be log-normally distributed, and the unconfined compression strength of each element was calculated from the mean unconfined compression strength μ_{qu}, the coefficient of variation COV_{qu}, the autocorrelation distance θ, and the normal random matrix X using the mid-point method (Matthies et al., 1997) with Cholesky decomposition (Kasama et al., 2008; Baecher and Christian, 2003).

The number of samples of unconfined compression strength shown in Figure 6.2 is 623,200, which means that the number of elements in the improvement area is 6232 elements per analysis section, and therefore the unconfined compression strength of all elements in the improvement area for 100 cases (sections) was sampled and compiled. The autocorrelation

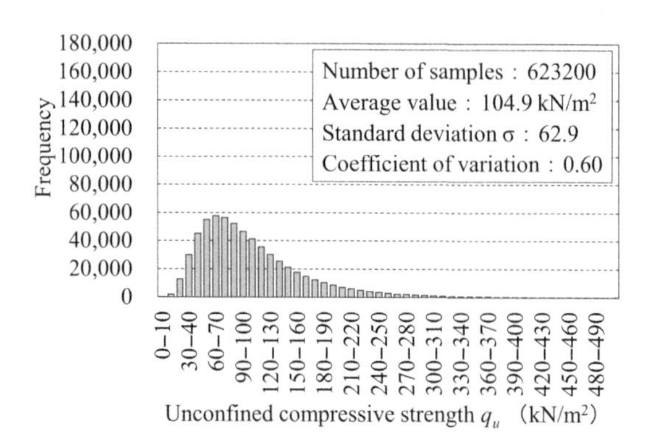

FIGURE 6.2 Distribution of unconfined compression strength by random field theory.

distance of the unconfined compression strength was set to 0.2 m in the vertical direction and 2 m in the horizontal direction based on the results of the unconfined compression strength obtained in the improved soil by chemical grouting at Airport A (Kasama et al., 2022). In this chapter, the deformation behavior of the runway due to dissipation of excess pore water pressure after liquefaction is probabilistically and statistically investigated by expressing the heterogeneity of shear strength using random field theory for an improved heterogeneous soil with some areas below the design base strength of 60 kN/m^2, as shown in Figure 6.2. In addition, it was also investigated that the results of the construction at other sites were lower than the design base strength, as shown in Figure 6.2 (CDIT, 2020). In the case of deep mix treated soils, which are also classified as improved soils, the quality check of heterogeneous improved soils is performed by allowing a certain amount of failure to occur relative to the design basis strength of the target improved body.

According to the *Technical Manual on Deep Mixing Methods for Ports and Airports*, it is acceptable to allow a failure rate (the percentage of strength that does not satisfy the design basis strength for a population of probability distributions obtained by assuming a normal distribution of unconfined compression strength) of up to about 15% in marine construction (CTRCF, 2018). Figure 6.3 shows an example of calculated unconfined compression strength for each element. Random field theory allows the use of different normal random matrices X to create a number of cases of soils with different spatial distributions of strength. Using the above method of evaluating the variation of shear strength in the improved soil by chemical injection, we performed deformation analysis with several variations of the compliance ratio and coefficient of variation to obtain the

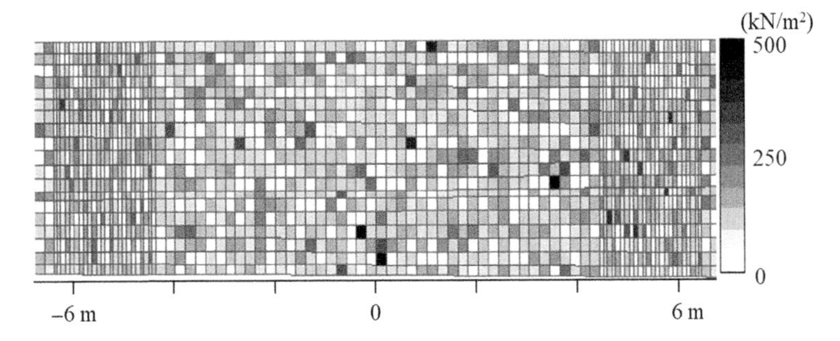

FIGURE 6.3 Example of calculation results of unconfined compression strength of each element.

relationship between the compliance ratio and the change in slope of the runway for each variation of the unconfined compression strength q_u.

The mesh spacing within the improvement area was basically divided into 0.2 m × 0.2 m meshes, taking into consideration that the autocorrelation distance of the unconfined compression strength in the vertical direction is about 0.2 to 0.3 m, which was confirmed in the ground investigation conducted by the authors (Kasama et al., 2022). However, the mesh around the aircraft load was divided more densely to ensure calculation accuracy, with the width of the mesh around the load being about half the width of the mesh around the aircraft load. As a result, the number of nodes and elements in the analytical mesh was 12,342 and 11,959, respectively. Figure 6.4 shows the analytical model diagram. Figure 6.5 shows the analytical mesh diagram.

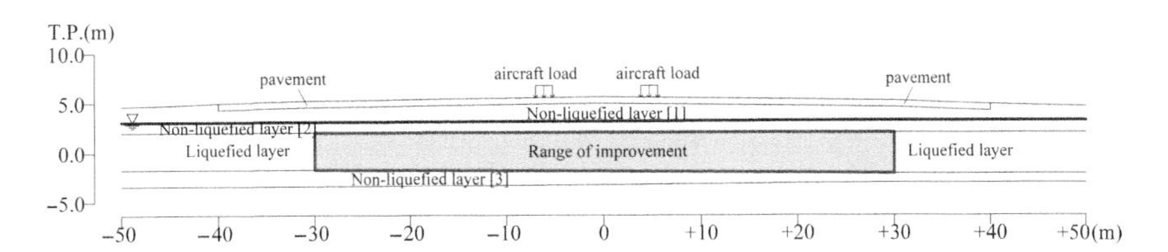

FIGURE 6.4 Analysis model.

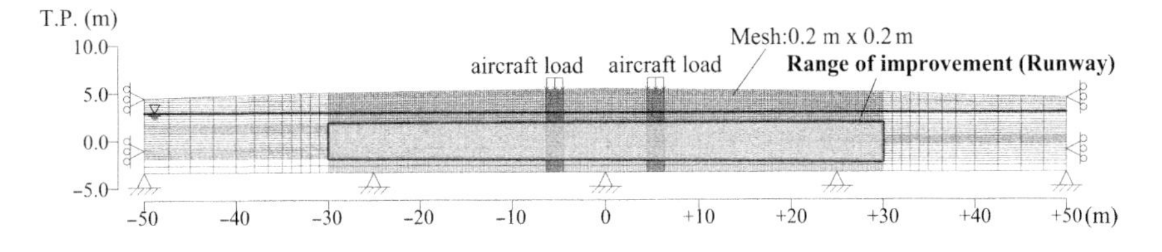

FIGURE 6.5 Analytical mesh diagram.

The total width of the pavement is 80 m. The width of the runway in the center is 60 m, and the width of the shoulder on the outside of the runway is 10 m. To simplify the analytical model, the runway and the shoulder were made of the same material. For the boundary conditions, the bottom of the analysis domain was assumed to have fixed displacements in the X and Y directions, and the sides of the analysis domain were assumed to have fixed displacements in the X direction and free displacements in the Y direction. The aircraft load was assumed to be the B777-9, the largest aircraft in the LA-1 fleet. Table 6.2 shows the load specifications for the B777-9.

Figure 6.6 shows the wheel arrangement of the B777-9 and the loads used in the deformation analysis. Table 6.3 shows the results of the aircraft

TABLE 6.2 Load Specifications for B777-9 Aircraft

Aircraft	Gross Mass (t)	Leg Load/Wheel Load (kN)	Ground Pressure (N/mm^2)	Ground Width (cm)
B777-9	352.4	1630/272	1.58	34.4

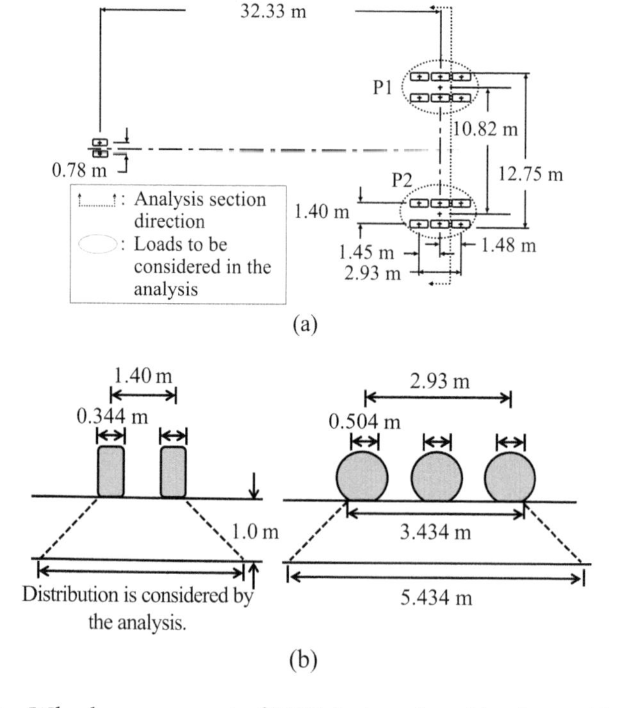

FIGURE 6.6 Wheel arrangement of B777-9 aircraft and loads considered in the analysis (a) wheel layout, (b) load distribution is considered by the analysis.

TABLE 6.3 Aircraft Load Condition

Aircraft		**B777-9**
Ground contact pressure		1.58 (N/mm^2)
Ground contact width		344 (mm)
Ground contact length		500 (mm)
Loads per a wheel P_0		$P_0 = 1.58 \times 344 \times 500 \div 1000 ≒ 272$ (kN)
Aircraft loads (impact load consideration) ※ 30% increase	P_1	$P_1 = 6 \times P_0/(1.400 + 0.344)/(2.930 + 0.504 + 1.000 \times 2) \times 1.3 ≒ 224$ (kN/m^2) ※ loading width on analysis model: 1.744 m
	P_2	$= P_1$

load settings. Since this deformation analysis is performed in the transverse direction of the runway, the loads (P1 and P2) were considered in the direction of the analysis section in Figure 6.6.

In the depth direction of the analytical cross-section, the distributed load was set considering the load dispersion up to the top of the roadbed (GL-1 m), because the load is dispersed in the roadbed. In the direction of the analytical cross-section, on the other hand, load dispersion is taken into account in the analysis, and therefore the range of load dispersion in the ground is not considered when setting the distributed loads. The aircraft loads were assumed to be the maximum possible loads, and therefore, load variations were not taken into account. In addition, a 30% surcharge (MLIT, 2019) of the static load set as the impact load due to the running of the aircraft was considered. Aircraft loads are treated as excess pore water pressure (hydraulic head). In this chapter, deformation verification considering aircraft loads was conducted immediately after the earthquake, so that excess pore water pressure equivalent to σ_v', in which aircraft loads are considered, is generated in the liquefaction elements. On the other hand, although excess pore water pressure is generated due to aircraft load, there is no significant volumetric strain for the improvement section because the excess pore water pressure ratio in this section is set to zero.

A list of analytical constants is shown in Table 6.4. The unit volume weight γ of the conventional layer is based on the results of wet density tests, the modulus of elasticity of the unimproved section is based on the results of PS tests, and the hydraulic conductivity of the non-pavement layers is based on the results of field or laboratory hydraulic tests. For other layers for which no test results were available, general values were used.

TABLE 6.4 List of Analytical Constants

Layer	Soil	Unit Weight		N-Value	Poisson Ratio v	Modulus of Elasticity E_0 (kN/m²)	Coefficient of Permeability k (cm/s)	Parameter	
		Wet γ_t (kN/m³)	Saturated γ_{sat} (kN/m³)					a	b
Pavement	—	22.5	22.5	—	0.35	414,000	1.0E-07	—	—
Non-liquefied layer [1]	Sandy	18.9	19.4	11	0.33	194,000	9.5E-03	—	—
Non-liquefied layer [2]	Clay	16.5	16.6	3	0.33	88,000	7.1E-05	—	—
Improvement area	Sandy	18.2	19.0	9	0.33	[a]	1.2E-06	9	4
Liquefied layer	Sandy	18.2	19.0	8	0.33	122,000	3.1E-03	9	4
Non-liquefied layer [3]	Clay	17.8	17.8	6	0.33	130,000	1.0E-06	—	—

[a] $E_0 = 122{,}000 + 12{,}200 \times q_u/q_{uck}$ (The unconfined compression strength, q_u, is set to the value of each element according to random field theory).

The modulus of elasticity of the improved area was set according to the ratio between the unconfined compression strength and the design base strength $q_{uck} = 60$ kN/m^2, assuming an increase of about 10% of that of the unimproved area (CDIT, 2020).

The constitutive equations for each material were set to linear elasticity for the pavement and the nonliquefied layer, and to exponential model for the improved area and the liquefied layer, which is the mechanical model used in FLIPDIS for consolidation deformation analysis. The pavement settled almost uniformly, and it was confirmed using the analysis results that no excessive tensile stresses were generated in the pavement. The parameters a and b of the exponential model were set to the general values ($a = 9$, $b = 4$) given by Morio and Katoh (2011). These values were obtained by simulating liquefaction experiments in which settlement associated with dissipation of excess pore water pressure was measured and reproducing the temporal variation of the time-settlement relationship.

Other required analytical constants are unit volume weight γ, Poisson's ratio ν, elastic modulus E_0, and hydraulic conductivity k. It should be noted that since this is a saturated analysis program, the permeability of the pavement and the nonliquefied layer [1], which is located below the groundwater table, is also considered.

The compliance rates analyzed were for the six cases shown in Table 6.5. As indicated by the authors (Kasama et al., 2022), the coefficients of variation of unconfined compression strength investigated for each block ranged from about 0.46 to 0.85. Based on this, the analysis was conducted

TABLE 6.5 List of Analysis Cases

Case		Precision Ratio (%)	Coefficient of Variation	Average Strength (kN/m²)
1		100.0	—	60.0
2	*a*	76.7	0.2	58.9
	b	76.7	1.0	148.6
3	*a*	72.0	0.2	68.4
	b	72.0	1.0	140.0
4	*a*	68.0	0.2	49.0
	b	68.0	1.0	112.3
5	*a*	61.4	0.2	43.2
	b	61.4	1.0	95.1
6	*a*	48.9	0.2	34.0
	b	48.9	1.0	66.0

for cases 0.2 and 1.0, assuming a lower to upper limit of the assumed coefficient of variation. In general, the variation of unconfined compression strength is expressed as a log-normal distribution, so there is a unique relationship between the mean value of unconfined compression strength, the goodness-of-fit ratio, and the coefficient of variation. The average strength in Table 6.5 is the average of the shear strength of the elements that do not meet the design basis strength, assuming zero due to liquefaction.

6.3 DEFORMATION VERIFICATION CONSIDERING SPATIAL HETEROGENEITY OF LIQUEFACTION CHARACTERISTICS

6.3.1 Dissipation of Excess Pore Water Pressure

First, the dissipation of excess pore water pressure is shown for Case-6a, which has the lowest fitness factor (48.9% fitness factor, 0.2 coefficient of variation) among all cases. In Case-6a, 100 different ground types with different spatial distributions of unconfined compression strength were created using random field theory, and deformation analysis was performed for each ground type. Figure 6.7 shows an example of unconfined compression strength distribution based on the random field theory. The black elements in the figure do not meet the design basis strength, and these elements were treated as liquefied elements with excess pore water pressure ratio of 1.0.

Figure 6.8 shows the frequency distribution of unconfined compression strength within the improvement area inputted in the deformation analysis of Case-6a. Since the compliance ratio of Case-6a is 48.9%, the percentage of liquefied elements that do not meet the design base strength is 51.1% for all elements in the improvement area. Figure 6.9 shows the distribution of the average effective confining pressure immediately after the earthquake.

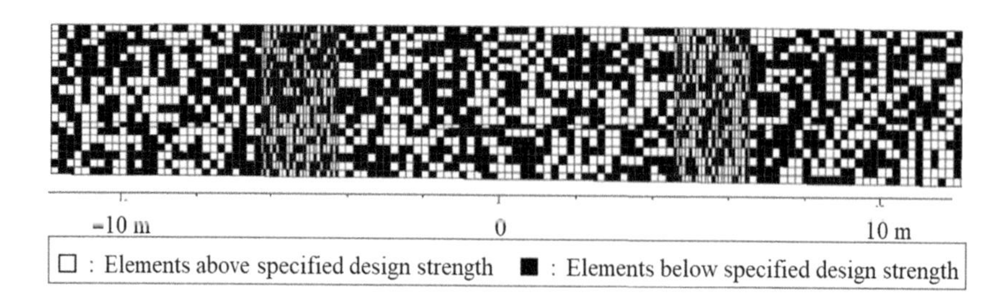

FIGURE 6.7 Distribution of unconfined compression strength by random field theory for Case-6a (precision ratio 48.9%, coefficient of variation 0.2).

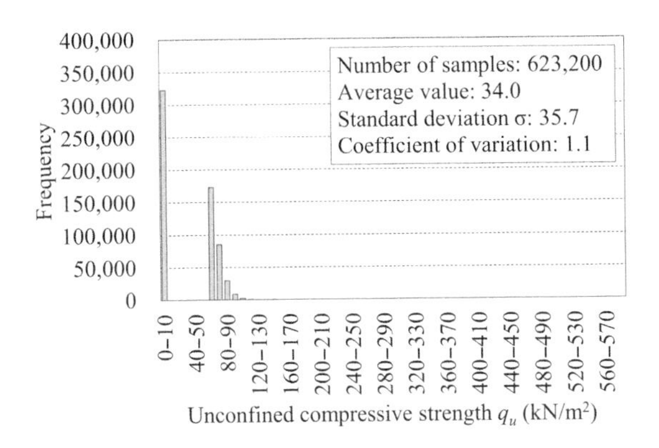

FIGURE 6.8 Distribution of unconfined compression strength during deformation analysis for Case-6a (precision ratio 48.9%, coefficient of variation 0.2).

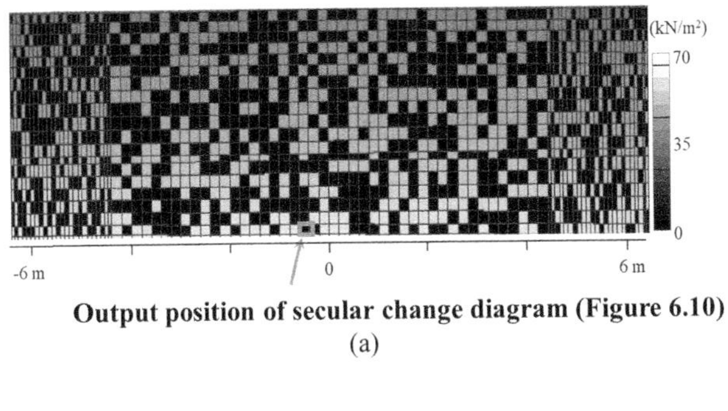

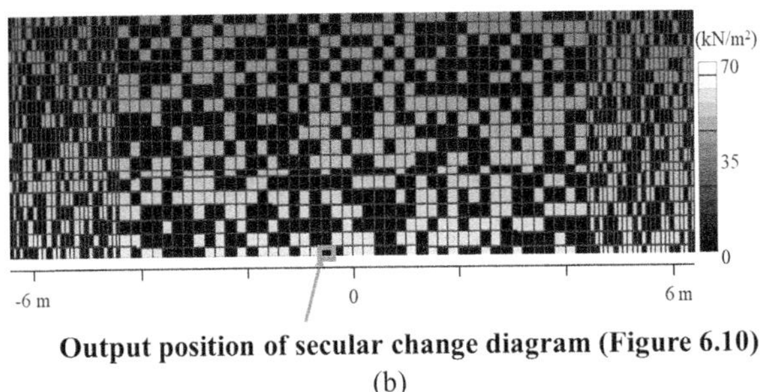

FIGURE 6.9 Distribution of average effective confining pressure for Case-6a (precision ratio 48.9%, coefficient of variation 0.2) (a) cases in which aircraft loads are not considered, (b) cases in which aircraft loads are considered.

Figure 6.9 shows the distribution of the average effective confining pressure immediately after the earthquake for the case without and with aircraft load.

First, an initial stress analysis was performed for the pre-earthquake condition without excess pore water pressure, and the effective stress of each element was analyzed. In a subsequent step, the excess pore water pressure of elements that reach an excess pore water pressure ratio of 1.0 due to full liquefaction is increased to a value equivalent to σ_v' to represent the ground immediately after liquefaction. Therefore, the case in which the aircraft load is taken into account is set so that excess pore water is generated due to the aircraft load compared to the case in which the aircraft load is not considered.

Figure 6.10 shows the distribution of the excess pore water pressure ratio. Figure 6.11 shows the time evolution of the excess pore water pressure near the bottom of the improvement area near the center of the analytical model, where the excess pore water pressure remains for the longest period of time. Figures 6.10 and 6.11 also show the case without and with aircraft load, respectively. Since the permeability of the improved area decreases due to the permeation of the chemical solution, it can be seen that only a few percent of the water pressure is dissipated immediately after the earthquake and three days after the earthquake. Therefore, in this chapter, the conditions for deformation analysis were set up for such

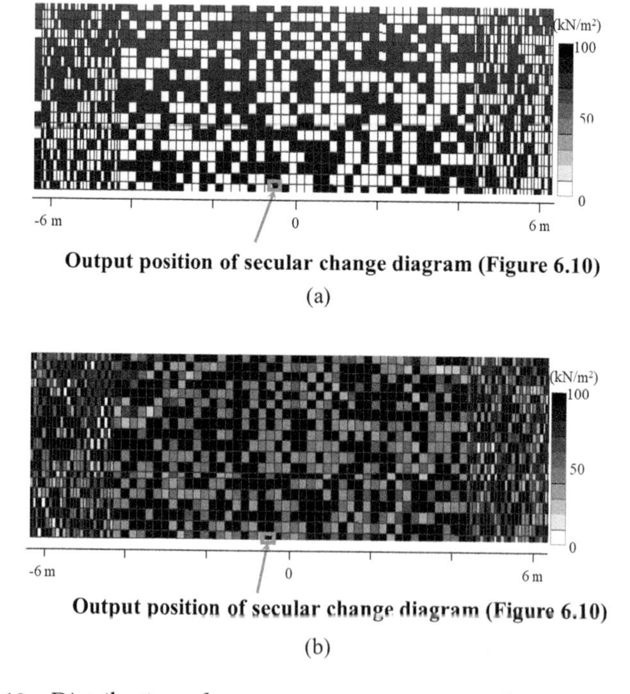

FIGURE 6.10 Distribution of excess pore water pressure for Case-6a (precision ratio 48.9%, coefficient of variation 0.2) (a) cases in which aircraft loads are not considered, (b) cases in which aircraft loads are considered.

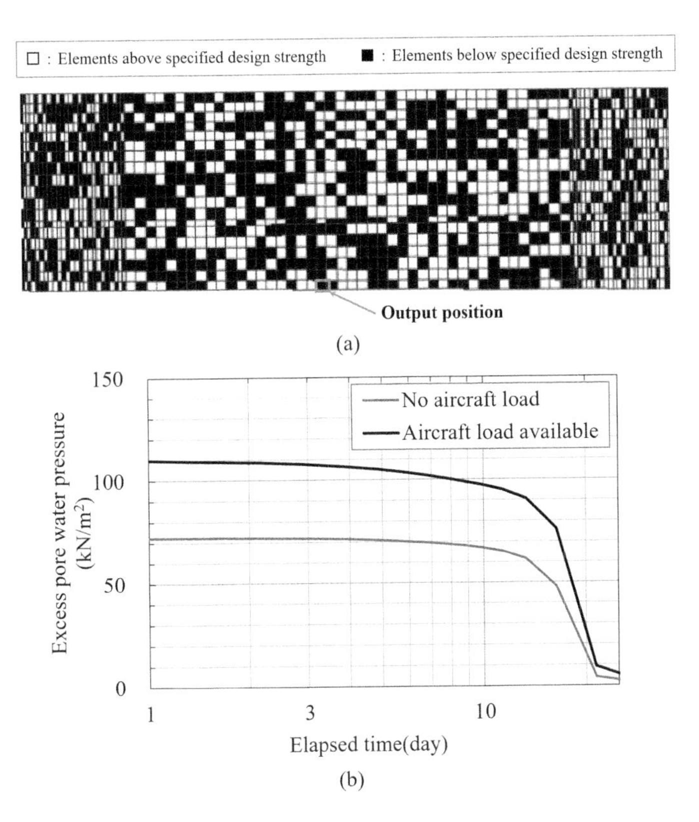

FIGURE 6.11 Excess pore water pressure dissipation in Case-6a (precision ratio 48.9%, coefficient of variation 0.2) (a) output position diagram, (b) secular change diagram.

heterogeneous improved ground, which requires time for the dissipation of excess pore water pressure based on a detailed understanding of the dissipation process of excess pore water pressure after the earthquake and the effect of the presence of aircraft load on the runway.

6.3.2 Status of Ground Surface Subsidence and Gradient Change

Next, the amount of ground surface subsidence and change in ground surface gradient are shown for Case-6a (precision ratio 48.9%, coefficient of variation 0.2). Figure 6.12 shows the distribution of surface settlement occurring at the runway and shoulder. Figure 6.12 shows the distribution of cases where the maximum or minimum ground settlement and gradient change immediately after the earthquake, three days after the earthquake, and after the dissipation of excess pore water pressure is found near the center of the runway (x = –20 m to +20 m section) for the case

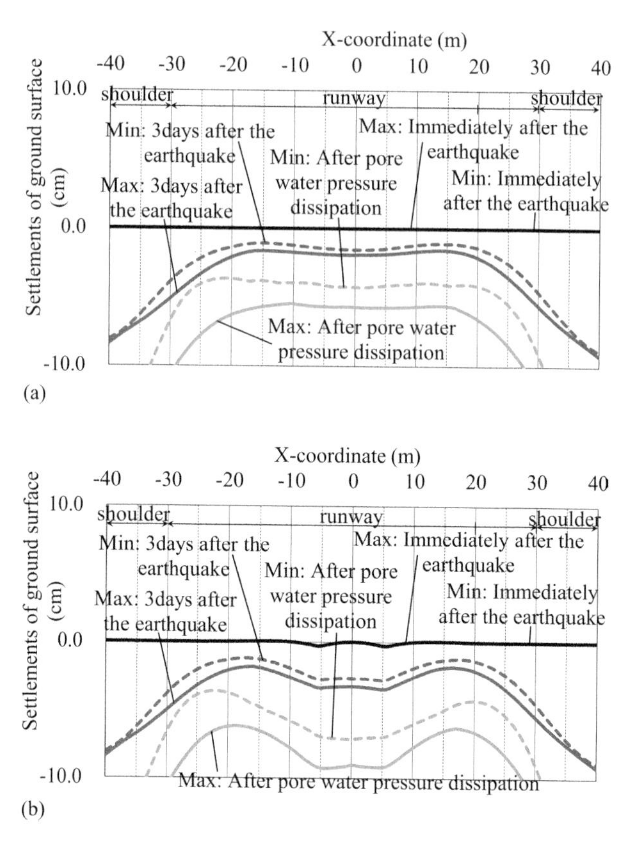

FIGURE 6.12 Surface settlement for Case-6a (precision ratio 48.9%, coefficient of variation 0.2).

without and with aircraft load, respectively, based on 100 Monte Carlo simulations.

Figure 6.13 shows the distribution of the gradient change of the ground surface in the central part of the runway (x = −20 m to +20 m section). Figure 6.13 shows the distribution of cases where the maximum or minimum ground settlement and gradient change immediately after the earthquake, three days after the earthquake, and after the dissipation of excess pore water pressure is found near the center of the runway (x = −20 m to +20 m section) for the cases without and with aircraft load, respectively, based on 100 Monte Carlo simulations. The denominator for calculating the gradient was set to 5 m, which is the measurement interval in the transverse direction during the maintenance of Airport A.

Figure 6.14 shows the frequency distribution of the maximum gradient change of the ground surface at the center of the runway. In Case-6a, the

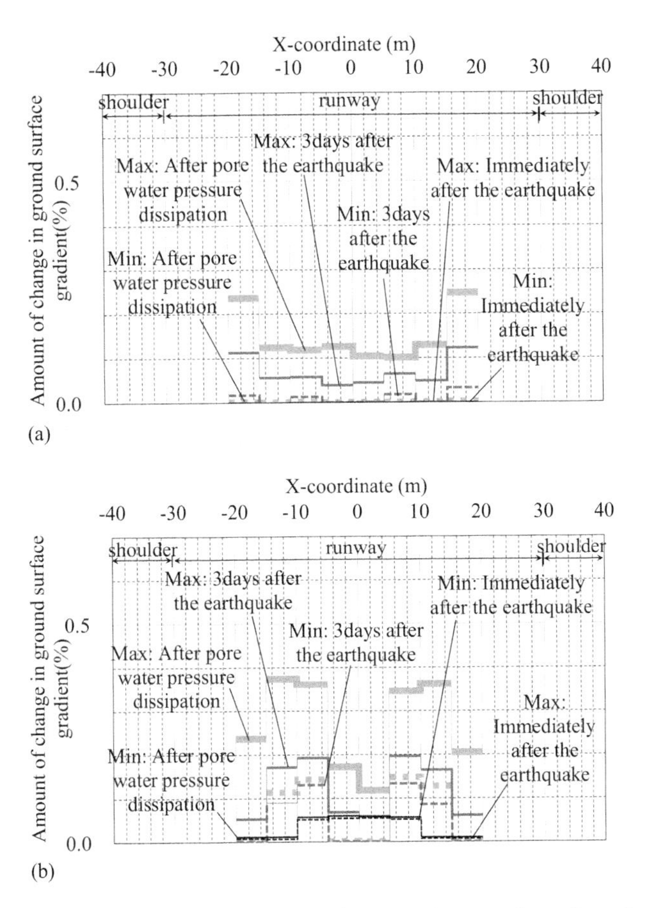

FIGURE 6.13 Distribution of gradient change of ground surface for Case-6a (precision ratio 48.9%, coefficient of variation 0.2).

maximum gradient change during the dissipation of excess pore water pressure exceeded the design limit (0.2%) in a part of the runway center, showing a variation of 0.05 to 0.25%. In the case where aircraft loads are considered, the maximum gradient change during the dissipation of excess pore water pressure exceeds the design limit (0.2%) over a wider range of values, showing a variation of 0.18 to 0.37%. These results indicate that the gradient change increases as the dissipation of excess pore water pressure progresses. Therefore, in the subsequent study, the verification focused on the amount of change in gradient when the excess pore water pressure dissipated completely. In this study, the influence of aircraft load on the gradient change was considered to be significant, and therefore, aircraft load was taken into account.

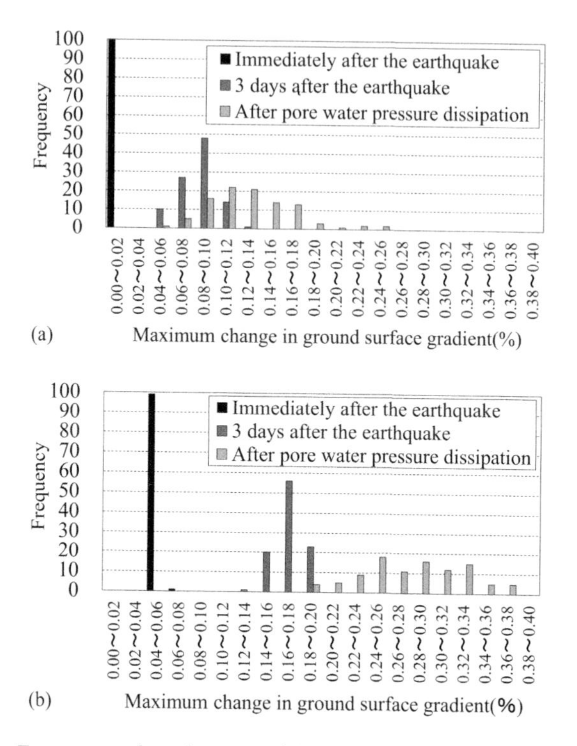

FIGURE 6.14 Frequency distribution of the maximum change in gradient of the ground surface for Case-6a (precision ratio 48.9%, coefficient of variation 0.2).

6.3.3 Confirmation of the Impact of Ground Heterogeneity

In order to confirm the effect of the variation of shear strength within the improvement area of the improved soil due to chemical grouting on the ground surface deformation, the deformations were compared for each typical compliance ratio. Figure 6.15 shows an example of the distribution of unconfined compression strength based on random field theory for each of the cases compared. The black-colored areas in the figure indicate elements that do not meet the design basis strength. These elements were treated as liquefied elements for which the excess pore water pressure ratio reaches 1.0. Specifically, the results of Case-1, in which the shear strength was not varied and the unconfined compression strength of the improved area was uniformly set to the design basis strength q_{uck} = 60 kN/m^2, or 100% compliance, were compared with those of Case-2 and Case-6a, in which the compliance was 76.7% (precision ratio 48.9%, coefficient of variation: 0.2). In Case-2—two cases, Case-2a with a coefficient of variation of 0.2 and Case-2b with a coefficient of variation of 1.0—were compared.

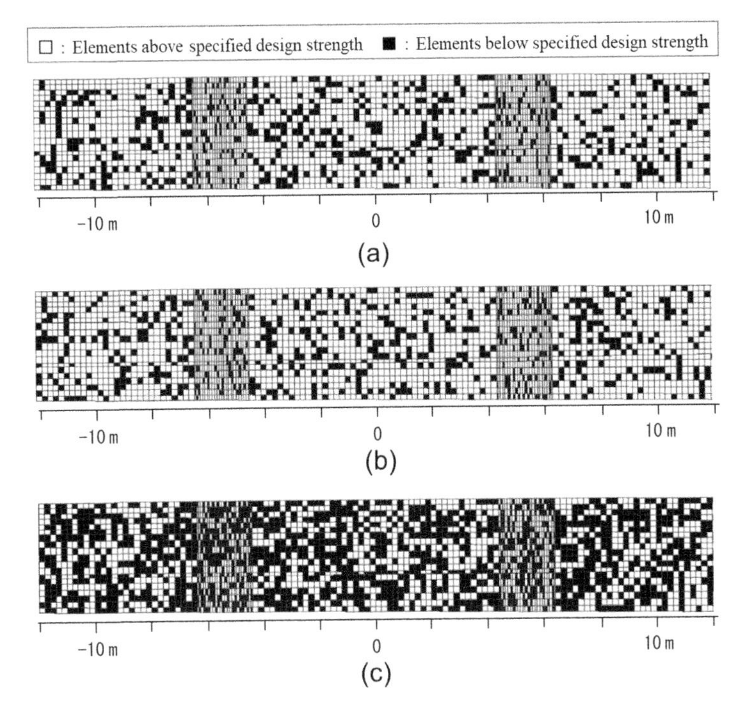

FIGURE 6.15 Distribution of unconfined compression strength by random field theory for each case (a) Case-2a (precision ratio 76.7%, coefficient of variation 0.1), (b) Case-2b (precision ratio 76.7%, coefficient of variation 0.1), (c) Case-6a (precision ratio 48.9%, coefficient of variation 0.2).

Figure 6.16 shows the frequency distribution of the unconfined compression strength entered in the deformation analysis for each case.

For example, the precision ratio for Case-2a is 76.7%, which means that the percentage of liquefied elements that do not meet the design basis strength is 23.3% for all elements in the improvement area.

Figure 6.17 shows an example of a mesh displacement diagram for each case. The deformation scale in Figure 6.17 is 20 times larger than the structure scale to make it easier to see the deformation within the improvement area. Compared to Case-1 (Figure 6.17(a)), where the settlement occurred around the area of aircraft load and there was no variation in shear strength, Case-6a (Figure 6.17(d)), where the precision ratio was the smallest, shows that the deformation extended to the area of improvement. There is no significant difference in the overall ground surface deformation shape between Case-2a (Figure 6.17(b)), in which the coefficient of variation is

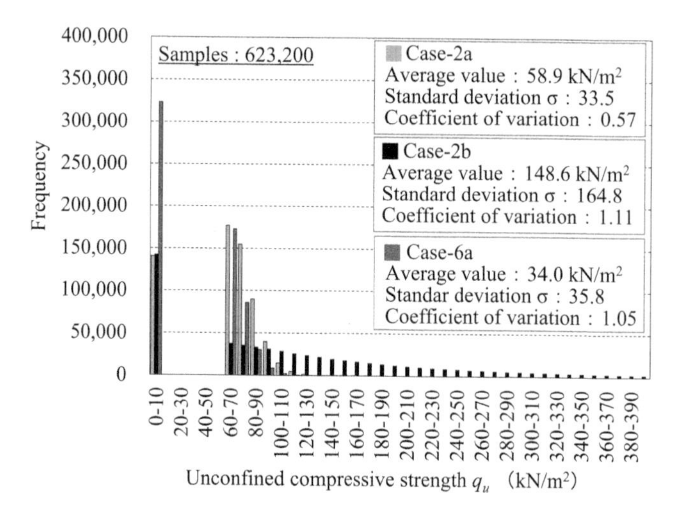

FIGURE 6.16 Frequency distribution of unconfined compression strength for each case of deformation analysis.

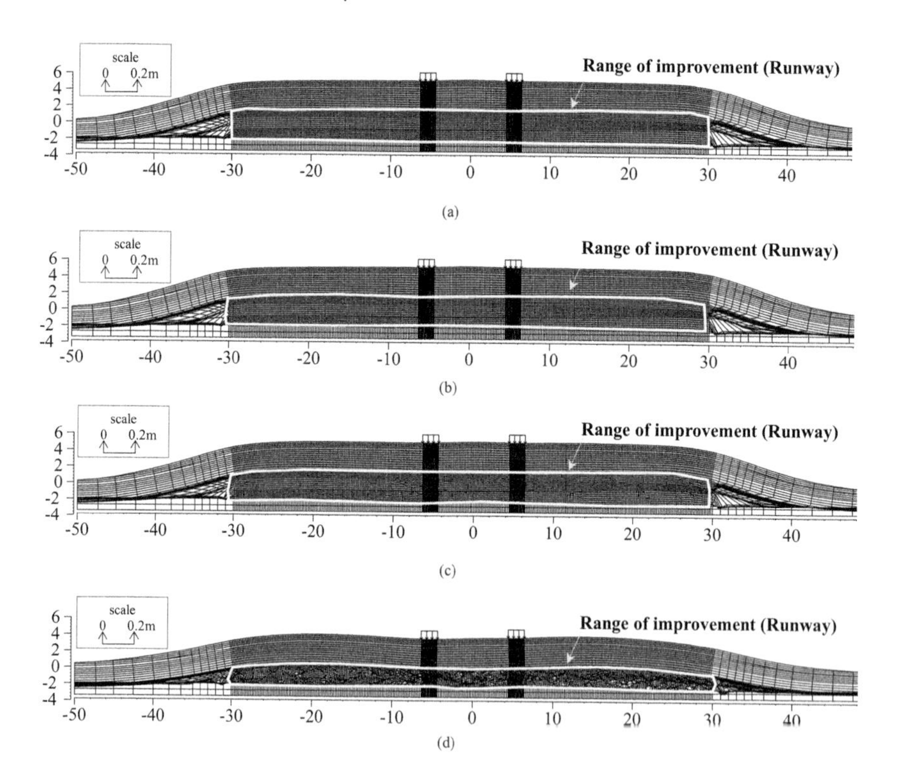

FIGURE 6.17 Mesh displacement diagram for each case (a) Case-1 (precision ratio 100%), (b) Case-2a (precision ratio 76.7%, coefficient of variation 0.2), (c) Case-2b (precision ratio 76.7%, coefficient of variation 1.0), (d) Case-6a (precision ratio 48.9%, coefficient of variation 0.2).

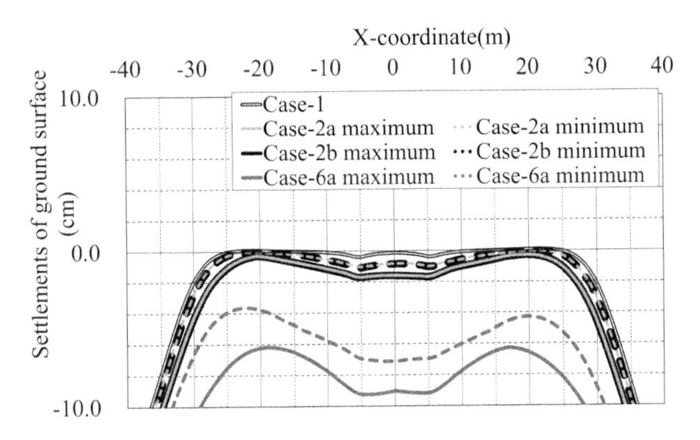

FIGURE 6.18 Distribution of ground surface settlement for each case.

set to 0.2 with a precision ratio of 76.7%, and Case-2b (Figure 6.17(c)), in which the coefficient of variation is set to 1.0 with the same precision ratio of 76.7%.

Figure 6.18 shows the distribution of cases with the largest or smallest settlement of the ground surface near the center of the runway ($x = -20$ m to +20 m section) during dissipation of excess pore water pressure, obtained from 100 Monte Carlo simulations for each case. The results show that Case-6a, which has the smallest precision ratio, has the largest deformation. The maximum settlement was about 9 cm near the aircraft load, 12 cm at the edge of the runway, and 19 cm at the shoulder. Volumetric strain in the area of improvement was about 0.1–2.5%, while volumetric strain outside the area of improvement was about 4–5%. Oshima et al. (2008) evaluated the deformation characteristics of solution-modified sand by reconsolidation tests after cyclic shearing. In their study, they reported that the final volumetric strain of unmodified sand was approximately equivalent to 2.5%, while the final volumetric strain of modified sand was less than 1%. The results of the present analysis showed a volumetric strain of about 0.1% at 100% precision ratio, which is consistent with the report by Oshima et al. (2008). Figure 6.19 shows the frequency distribution of the maximum surface settlement at the center of the runway ($x = -20$ m to +20 m) obtained from 100 Monte Carlo simulations for Case-2a with the coefficient of variation set to 2.0 and Case-2b with the coefficient of variation set to 1.0. There is no clear difference in the variation of settlement between the two cases.

Figure 6.20 shows the distribution of the maximum and minimum values of the change in surface gradient during excess pore water pressure

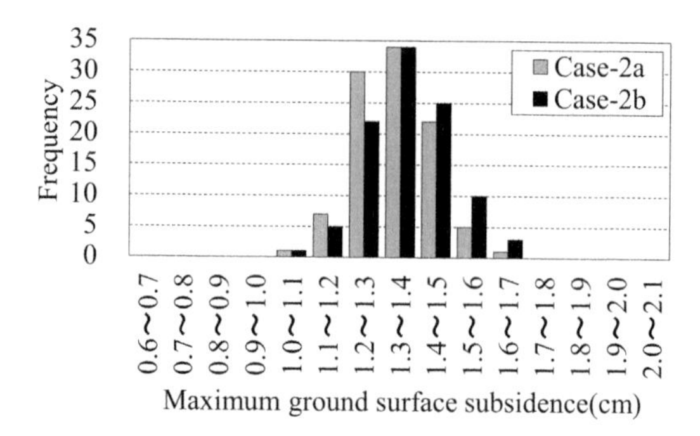

FIGURE 6.19 Frequency distribution of maximum surface settlement for Case-2 (precision ratio 76.7%, coefficients of variation 0.2 and 1.0).

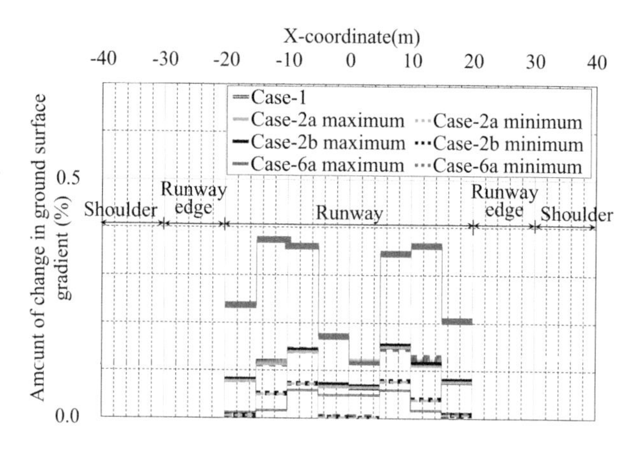

FIGURE 6.20 Distribution of ground surface gradient change for each case.

dissipation obtained from 100 Monte Carlo simulations for each case near the center of the runway (x = −20 m to +20 m section). When assuming a B777-9 aircraft, the maximum cross-sectional gradient of the runway is specified as 1.5% because of the aircraft's wingspan and Code E. In the following discussion, the current runway cross slope is assumed to be 1.3% and the limit value of the slope change is 0.2%.

Figure 6.21 shows the frequency distribution of the maximum change in gradient of the ground surface in the central part of the runway (x = −20 m to +20 m section), obtained from 100 Monte Carlo simulations. These results show that Case-6a, which has the smallest precision ratio, has the largest gradient change, and the maximum gradient change exceeds the design limit value (0.2%), indicating a variation of 0.18 to 0.37%.

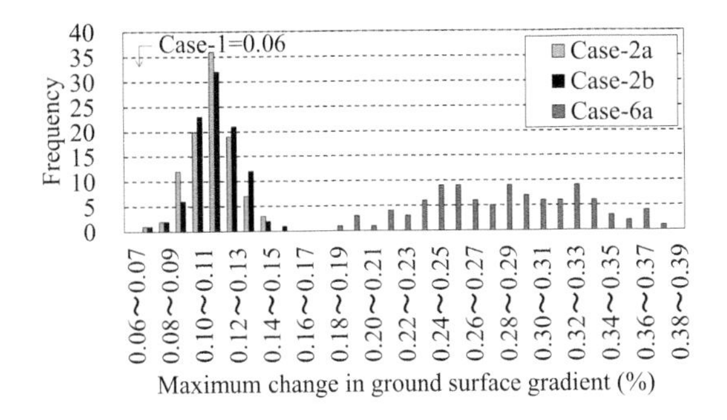

FIGURE 6.21 Frequency distribution of the maximum change in gradient of the ground surface for each case.

6.4 PERFORMANCE-DEFINED DEFORMATION VERIFICATION METHOD

As indicated by the authors (Kasama et al., 2022), the precision ratio of unconfined compression strengths investigated in each block ranged from about 62 to 97%, with coefficients of variation ranging from 0.46 to 0.85. Based on these results, a deformation analysis was conducted for six cases of precision ratio, varying the coefficient of variation between 0.2 and 1.0, using a method to evaluate the variability of shear strength in the range of improvement of the improved soil by chemical grouting. From the results, the relationship between the maximum settlement and gradient change of the runway and the precision ratio for each variation of shear strength (unconfined compression strength q_u in the improved ground) was determined.

Figure 6.22 shows the relationship between the maximum settlement of the runway section and the precision ratio. Figure 6.22 shows the maximum settlement for a range of assumed coefficients of variation of unconfined compression strength of the ground (0.2 and 1.0). Figure 6.23 shows the relationship between the gradient change of the runway section and the precision ratio. In Figure 6.23, the results of the slope change for the assumed range of coefficients of variation of the shear strength of the ground (0.2 and 1.0) are shown. These results show that there is no significant difference between the two cases for the same precision ratio, although the maximum settlement and gradient change are slightly smaller for the case with a coefficient of variation of 1.0, which has a larger variation. This indicates that the influence on the deformation of the ground surface depends more on the precision ratio, which is the percentage of the design

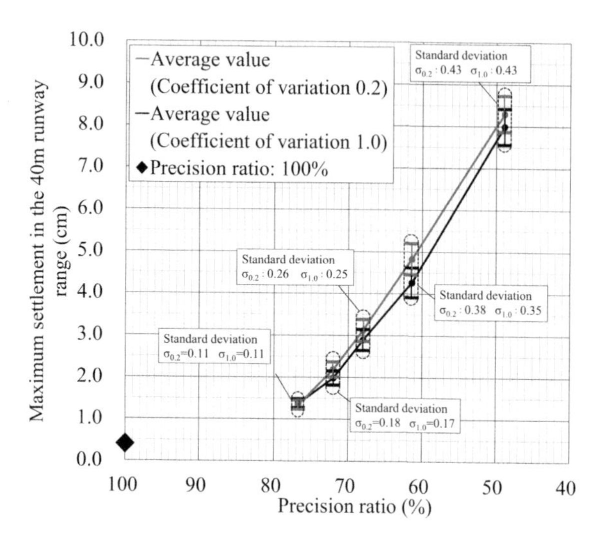

FIGURE 6.22 Relation between maximum settlement and precision ratio.

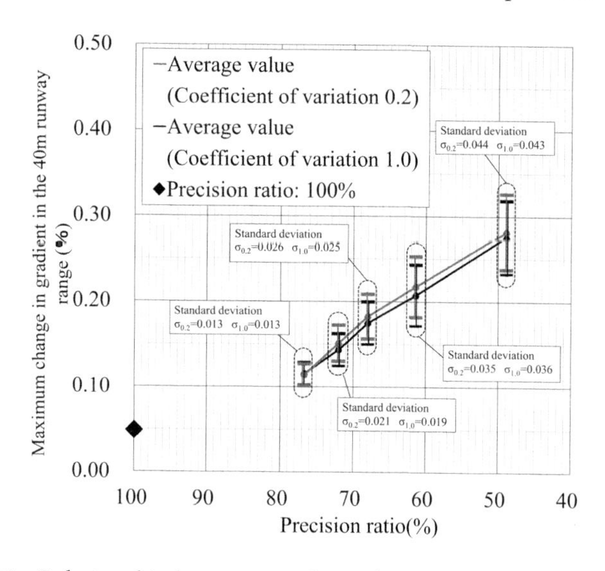

FIGURE 6.23 Relationship between gradient change and precision ratio.

strength satisfied, than on the spatial variation of the factors satisfying the design strength.

Using these results, it is possible to easily estimate the approximate value of the change in gradient by determining the unconfined compression strength of the post-improvement soil after chemical grouting without conducting a detailed deformation analysis. For example, as shown in Figure 6.24, if the runway cross gradient before the earthquake is 1.3%,

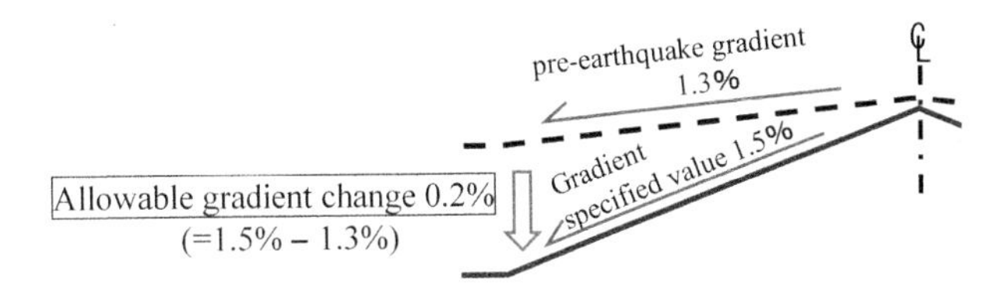

FIGURE 6.24 Diagram of allowable gradient change.

the gradient change is allowed to be within 0.2% of the specified runway cross gradient of 1.5%. Therefore, as shown in Figure 6.23, if the precision ratio is generally greater than 70%, the gradient change can be evaluated to be smaller than the allowable value of 0.2%, regardless of the variation (coefficient of variation) (Figure 6.24).

6.5 CONCLUSIONS

In this chapter, deformation verification due to dissipation of excess pore water pressure after an earthquake was conducted for the purpose of verifying the flatness of an airport runway based on the performance specifications for a ground improved by chemical grouting to prevent liquefaction of sandy soil, taking into account the heterogeneity of shear strength. As a result, the following points were found:

1) The maximum settlement of the airport runway caused by insufficient strength of the improved ground increases rapidly around 80% precision ratio, with a settlement of approximately 8.0 cm at 50% precision ratio. The difference in the maximum settlement due to differences in strength (coefficient of variation) was about 0 to 0.5 cm (standard deviation of 0 to 0.03 cm), and it was confirmed that there was no significant difference. This indicates that the influence on the deformation of the ground surface depends more on the precision ratio, which is the percentage of the design strength satisfied, than on the spatial variation of the factors satisfying the design strength.

2) The maximum change in gradient of the airport runway after liquefaction increased linearly (mean 0.12 to 0.28%, standard deviation ±0.013 to ±0.044%) in the range of 76.7 to 48.9% of the precision ratio. Using these results, it is possible to estimate the approximate value of the change in gradient by determining the unconfined compression strength of thc improved soil after chemical grouting.

REFERENCES

Baecher, G. B. and Christian, J. T., 2003. *Reliability and Statistics in Geotechinical Engineering*. John Wiley & Sons.

Biot, M. A., 1941. General theory of three-dimensional consolidation. *Journal of Applied Physics* 12, 155–164.

Civil Aviation Bureau, Ministry of Land, Infrastructure, Transport and Tourism, 2019. *Design Guidelines and Examples for Airport Civil Engineering Facilities (Seismic Design).*

Coastal Development Institute of Technology, 2020. *Technical Manual of Permeation Grouting Method*, p. 55.

Coastal Technology Research Center Foundation, 2018. *Coastal Engineering Library No. 29 Technical Manual for Deep Mixing Methods for Offshore Construction* (Revised Edition), pp. 38–39.

Hatanaka, J. and Murono T., 2002. Nonlinear seismic response characteristics of surface soils considering variation of ground constants, *The 11th Japan Earthquake Engineering Symposium*, Tokyo, Japan, pp. 769–774.

Honjo, Y. and Otake, Y., 2012. A simplified scheme to evaluate spatial variability and statistical estimation error of local average of geotechnical parameters in reliability analysis. *Journal of Japan Society of Civil Engineers* 68, 41–55.

Iai, S., Matsunaga, Y., and Kameoka, T.: Strain space plasticity model for cyclic mobility. *Soils and Foundations* 32(2), 1–15, 1992.

Juang, H. C., Yang, H. S., and Yuan, H., 2005. Model uncertainty of shear wave velocity-based method for liquefaction potential evaluation. *Journal of Geotechnical and Geoenviromental Engineering, ASCE* 131(10), 1274–1282.

Kasama, K., Nagayama, T., Hamaguchi, N., Sugimura, Y., Fujii, T., Kaneko, T., and Zen, K., 2022. Performance-based evaluation for the bearing capacity of ground improved by permeation grouting method. *Journal of Japan Society of Civil Engineers* 78, 45–59.

Kasama, K., Zen, K., Chen, G., Kentaro, H., 2010. Shaking table test on the dynamic property of solidified ground considering the spatial locality of liquefaction. *Japanese Geotechnical Journal* 5, 241–250.

Kasama, K., Zen, K., and Whittle, A. J., 2008. Bearing capacity characteristics of cohesive land slabs using stochastic numerical limit analysis. *Journal of Applied Mechanics* 11, 291–298.

Kataoka, N., Kasama, K., Zen, K., and Chen, G., 2011. The liquefaction risk analysis of cement-treated sandy ground considering the spatial variability of soil strength. *Journal of Japan Society of Civil Engineers* 68, 68–83.

Matthies, H. G., Brenner, C. E., Bucher, C. G., and Guedes Soares, C., 1997. Uncertainties in probabilistic numerical analysis of structures and solids - Stochastic finite elements. *Structural Safety* 19(3), 283–336.

Miyata, M., Iai, S., and Ichii, K., 1998. Estimation of differential settlements due to liquefaction, *Report of the Port and Harbour Research Institute* 908, 1–24.

Morio, S. and Kato, Y., 2011. *On the Application of Biot's Consolidation Theory To The Post-Liquefaction Process Involving Water Film Generation*. Bulletin of National Institute of Technology Maizuru College, pp. 19–26.

Nakamura, S., Sawada, S., and Matsumoto, T., 2007. Ratio of nonlinear seismic response properties to the modeled dimension of heterogeneous spatial distribution of geotechnical properties. *Collection of Papers of the Japan Society of Civil Engineers C* 63(3), 711–724.

Ohshima, T., Kazama, M., Sendo, N., Kawamura, K., and Hayashi, K., 2008. Evaluation of liquefaction resistance and deformation characteristics after cyclic shear of solution chemical improved sand. *Collection of Papers of the Japan Society of Civil Engineers C* 64(4), 732–745.

Oka, F., Yashima, A., Shibata, T., Kato, M., and Uzuoka, R. 1994. FEM-FDM coupled liquefaction analysis of a porous soil using an elasto-plastic model. *Applied Scientific Research* 52, 209–245.

Otake, Y. and Honjo, Y., 2012. A practical geotechnical reliability based design employing response surface -Seismic design of irrigation channel on liquefiable ground. *Journal of Japan Society of Civil Engineers* 68, 68–83.

Popescu, R., Prevost, J. H., and Deodatis, G., 2005. 3D effects in seismic liquefaction of stochastically variable soil deposits. *Geotechnique* 55(1), 21–31. 209–245, 1994.

Shnabel, P. B., Lysmeer, J., and Seed, H. B., 1972. *SHAKE, A Computer Program for Earthquake Response Analysis of Horizontally Layered Sites*. University of California at Berkeley, EERC72-12.

Suetomi, I., Sawada, S., Yoshida, N., Toki K., 2000. Relation between shear strength of soil and upper limit of earthquake ground motion. *Journal of Japan Society of Civil Engineers* I-52, 195–206.

Sugano, T., Nakazawa, H., 2009. Experimental study on countermeasures for liquefaction subjected to full-scale airport facilities, Technical Note of the Port and Airport Research Institute, No. 1195, pp. 333–340.

Tsutida, T. and Ono, K., 1988. Prediction of unequal settlement by numerical simulation and its application to airport pavement design. *Report of the Port and Harbour Research Institute* 27(4), 123–200.

Uzuoka, R., Mihara, M., Yashima, A., and Kawakami, T., 1997. An analysis of lateral spreading of liquefied subsoil based on Bingham model. *Numerical Models in Geomechanics* IV, 685–690.

Yasuda, S., Yoshida, N., Adachi, K., Kiku, H., Gose, S., and Masuda, T., 1999. A simplified practical method for evaluating liquefaction-induced flow. *Journal of Japan Society of Civil Engineers* III-49, 71–89.

Zen, K., Yamazaki, H., and Sato, Y., 1990. Strength and deformation characteristics of cement treated sands used for premixing method. *Report of the Port and Harbour Research Institute* 29, 85–118.

CHAPTER 7

Performance Evaluation of Airport Runways after an Earthquake

Tomoyuki Kaneko

7.1 INTRODUCTION

The ground improved by the chemical grouting has spatial heterogeneity in the material constants due to the heterogeneity of chemical infiltration and the heterogeneity of the soil. This heterogeneity of the improved soil is expected to affect the bearing capacity and deformation of the improved soil at the time of design, and an evaluation method for bearing capacity and deformation based on performance specifications that take the heterogeneity of the soil into account is practically required. However, no previous studies have verified the heterogeneity of previously improved ground in detail using actual data, and no previous studies have presented methods for evaluating bearing capacity and deformation based on performance specifications that assume variability.

Therefore, in Chapter 5, the heterogeneity of shear strength of improved ground was verified in detail using actual data for improved ground treated to prevent liquefaction at airports. Shear strength was expressed using random field theory and Monte Carlo simulation using the finite element method, and shear strength reduction method to the bearing capacity analysis considering the strength heterogeneity was performed

DOI: 10.1201/9781032670133-7

by Monte Carlo simulation using the finite element method and shear strength reduction method.

In Chapter 6, we analyzed the deformation behavior of the ground due to dissipation of excess pore water pressure after liquefaction by expressing the shear strength in random field theory and by simply modeling the liquefaction of the ground during an earthquake for the same improved ground—and also discussed the probabilistic and statistical effects of heterogeneity of the improved ground on the deformation of the airport runway after an earthquake. Based on the verification results in Chapters 5 and 6, a new performance evaluation method based on performance specifications is proposed in this chapter for the bearing capacity and flatness required for airport runways after earthquakes by developing a chart to confirm the precision ratio that satisfies both bearing capacity and runway gradient performance after an earthquake. We also describe a case study in which the proposed performance evaluation method was applied to the actual ground surface of Airport A, which was liquefied by chemical grouting. In this chapter, we propose a more versatile performance evaluation method for blocks whose unconfined compression strength distribution does not conform to a lognormal distribution—and for blocks for which the performance evaluation by chart plotting does not satisfy the permissible values.

7.2 PROPOSAL FOR A PERFORMANCE EVALUATION METHOD FOR AIRPORT RUNWAYS AFTER AN EARTHQUAKE

7.2.1 Charting the Bearing Capacity Safety Factor and Deformation of the Ground after an Earthquake

In Chapter 5, the authors analyzed the bearing capacity of liquefaction-resistant previously consolidated soil by expressing the shear strength in random field theory and considering the heterogeneity of strength by Monte Carlo simulation using the finite element method and the shear strength reduction method. In addition, in Chapter 6, the same improved soil is analyzed for the performance of an airport runway after an earthquake by expressing the shear strength in random field theory and by modeling the liquefaction of the soil during an earthquake in a simplified manner. Using the results of these analyses, a chart was prepared to determine the precision ratio and variation of the unconfined compression strength of the soil improved by chemical grouting, which enables a

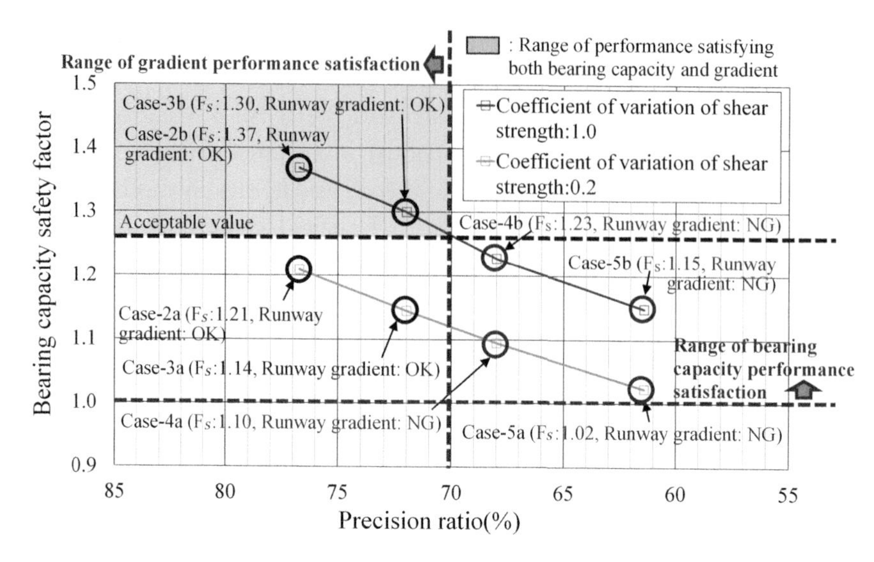

FIGURE 7.1 Chart for confirmation of precision rate for bearing capacity safety and deformation.

simple determination of whether or not the required runway gradient and other flatness are secured after an earthquake.

Figure 7.1 shows the relationship between the gradient verification results and the precision ratio in the 40 m range in the center of the runway, which was studied in Chapter 6, in addition to the "relationship between the level of safety of bearing capacity and the precision ratio" shown in Chapter 5.

The figure shows that if the shear strength precision ratio is greater than 60%, the allowable gradient change (0.2%) is not satisfied, but the bearing capacity safety (1.0) is satisfied. If the precision ratio is greater than 70%, both the bearing capacity safety (1.0) and the allowable gradient change (0.2%) will be satisfied. If the bearing capacity safety factor is greater than 1.3, both the bearing capacity safety factor and the gradient will satisfy the allowable values, indicating that the relationship is such that all the performance of the runway after a level 2 earthquake motion can be satisfied. From the chart diagram in Figure 7.1, it is possible to confirm that the precision ratio is sufficient to satisfy the performance of both bearing capacity and runway gradient after an earthquake, regardless of the coefficient of variation of unconfined compression strength (0.2 to 1.0) of the improved treated soil. The bearing capacity and deformation verification based on this chart diagram uses the coefficient of variation of unconfined compression strength and fitness factor obtained from the posterior investigation of the improved ground and plotted on the design line shown in Figure 7.1 to verify whether the ground satisfies the bearing capacity safety and the allowable gradient of the runway.

However, the "relationship between the degree of safety of bearing capacity and the precision ratio" is a relationship obtained based on the assumption that the distribution of unconfined compression strength conforms to a logarithmic normal distribution. The stratification conditions of the ground and the cross-sectional shape of the runway were obtained based on the results of the geotechnical investigation of Airport A, which was the subject of this study. Therefore, when using this relationship to evaluate the performance of other airports, it is necessary to take into account differences in local conditions such as aircraft loads to be considered, existing runway gradient, and ground conditions at the target airport. In cases where the local conditions differ significantly from those of Airport A, or where the precision ratio is extremely low, or where the distribution of unconfined compression strength does not conform to a lognormal distribution—as in the case of Block d of Airport A, the subject of this chapter—it is necessary to perform bearing capacity and deformation verification based on detailed modeling of the ground conditions under consideration, as was done in this chapter. Based on the above, a flow diagram of the method for evaluating the performance of airport runways after an earthquake is shown in Figure 7.2.

7.2.2 Performance Verification of Airport a Based on Chart Drawings

Performance evaluation was conducted on the actual ground of Airport A, which was liquefied by the chemical grouting, a typical method among the chemical grouting methods, based on the chart diagram created in Section 7.2.1, Charting the Bearing Capacity Safety Factor and Deformation of the Ground after an Earthquake. The evaluation results are shown in Table 7.1 and Figure 7.3.

By plotting the coefficient of variation (COV) of the shear strength of the ground based on the unconfined compression strength obtained from the posterior results of each block (coefficient of variation of shear strength of the soil: COV = 0.2, 0.4, 0.6, 0.8, or 1.0) on the design line that is closest and on the safe side, it was verified whether or not the value satisfied the allowable value of bearing capacity safety factor.

From the table, for Block b and Block e, the evaluation based on the chart diagram resulted in a NG judgment (i.e., the performance of the

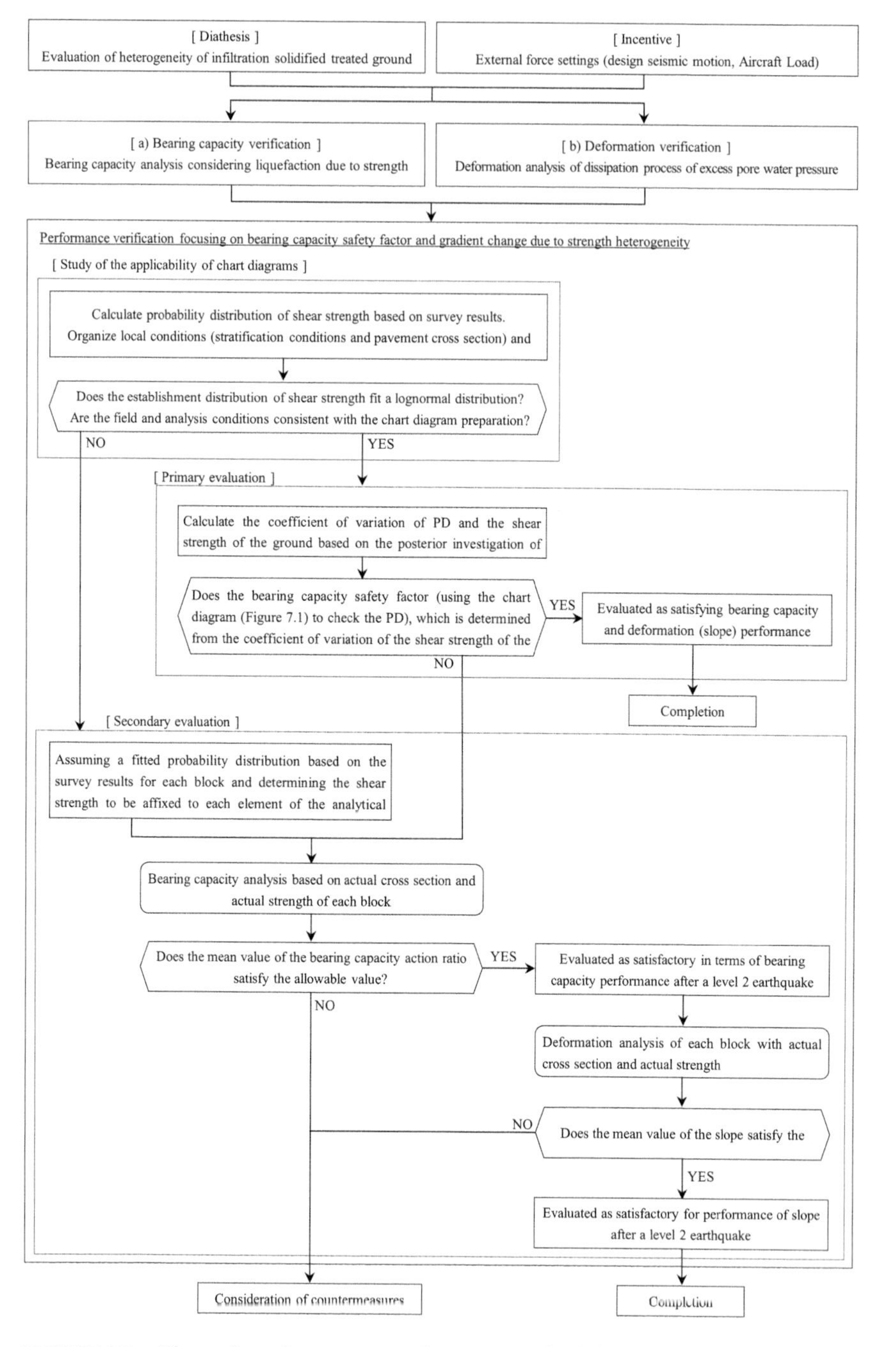

FIGURE 7.2 Flow of performance evaluation method for airport runways after an earthquake.

TABLE 7.1 Evaluation Results from the Primary Evaluation

Block	Number of Survey Sites	Coefficient of Variation	Precision Ratio[a] (%)	Judgment Result
a	4	0.590	91	OK
b	4	0.774	77	**NG**
c	4	0.552	94	OK
d	4	0.654	86	**OK**[b]
e	4	0.846	62	**NG**
f	4	0.638	83	OK
g	4	0.540	94	OK
h	4	0.544	89	OK
i	4	0.714	80	OK
j	4	0.541	95	OK
k	3	0.586	92	OK
l	3	0.466	97	OK
m	3	0.616	83	OK

Notes:

[a] The probability distribution obtained by assuming a lognormal distribution for the posterior results is used as the population and the percentage of the distribution that satisfies the design basis strength for that population.

[b] Unconfined compression strength obtained from the posterior results did not confirm the goodness of fit of the lognormal distribution, so a secondary evaluation is needed.

bearing capacity and deformation was not secured). Therefore, a secondary evaluation, which is a more detailed evaluation by individual cross-sections, is necessary. However, in the primary evaluation, when all data within the improvement range was aggregated, it was confirmed that the data fit the lognormal distribution, but when it was checked for individual blocks, only Block d was not confirmed to fit the lognormal distribution. Therefore, it was decided to check the Block d in more detail in the secondary evaluation. In the secondary evaluation, two types of evaluations were conducted: One using the intensity distribution obtained in the survey of Block d as it is, and the other assuming a lognormal distribution.

7.2.3 What to Do When Variation Is Not Normally Distributed

7.2.3.1 Consideration of Heterogeneity of Ground Improved by Chemical Grouting

In general, the distribution of measured unconfined compression strength in chemical grouting improved soils tends to fit a lognormal distribution. The authors conducted a Monte Carlo simulation of bearing capacity analysis and deformation verification considering strength heterogeneity

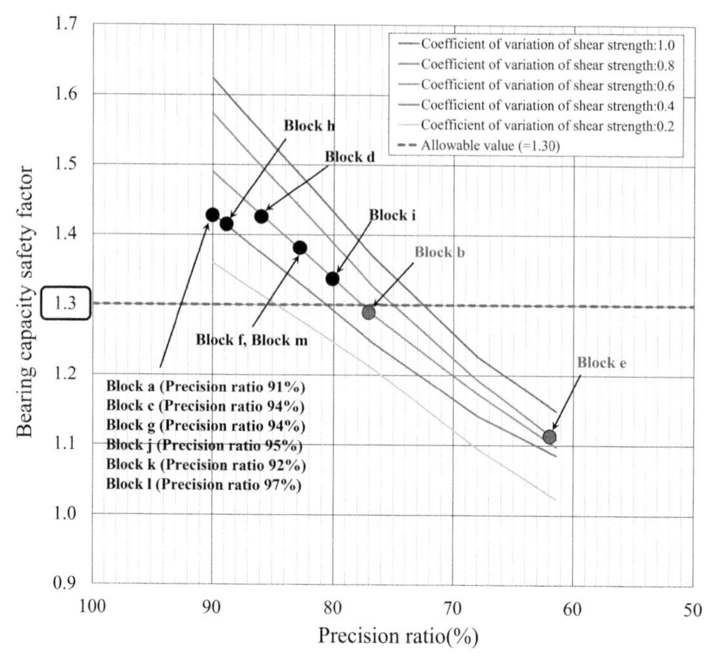

*Coefficient of variation greater than 0.4 to 0.6: Plotted on the line with a coefficient of variation of 0.4 in Figure 7.3.
Coefficient of variation greater than 0.6 to 0.8: Plotted on the line with a coefficient of variation of 0.6 in Figure 7.3.
Coefficient of variation greater than 0.8 to 1.0: Plotted on the line with a coefficient of variation of 0.8 in Figure 7.3.

FIGURE 7.3 Evaluation results based on chart diagram.

after confirming with a chi-square test that the distribution of measured unconfined compression strength in ground improved by chemical grouting conformed to a lognormal distribution at a significance level of 5%. On the other hand, in some construction areas, the distribution of measured unconfined compression strength of the improved ground by chemical grouting did not conform to a lognormal distribution. Therefore, in this chapter, we assume that the distribution of unconfined compression strength does not conform to the lognormal distribution to extend the range of applicability of the proposed verification method. The heterogeneity of the unconfined compression strength of the improved soil was expressed by using the bootstrap method (Davison, 1997) for the distribution of measured unconfined compression strength.

The bootstrap method is a resampling technique formulated by Bradley Efron of Stanford University. It is a nonparametric method that automatically obtains the probability distribution of an estimator from the data itself and estimates the standard deviation and confidence interval without assuming a specific probability distribution such as a normal distribution by repeating a random sampling from the sample multiple times. In this chapter, the unconfined compression strength of each element is calculated using the same method to account for the heterogeneity of the unconfined compression strength, and the bearing capacity analysis is performed by Monte Carlo simulation using the finite element method and shear strength reduction method—and the deformation analysis associated with the dissipation of excess pore water pressure after liquefaction is performed.

7.2.3.2 Airport A Overview and Improved Ground with Chemical Grouting

Figure 7.4 shows a plan view of the area improved by chemical grouting at Airport A.

The construction work was carried out by dividing the area into 13 blocks. Airport A is located on a plain that forms an urban area. Around the airport, hills and terraces are formed, and the area is located on a plain where a river flows between these hills and terraces. In the area under investigation, the stratigraphic structure consists primarily of a 2 m layer of old fill (layer B), a 1 m layer of low permeability clay, and a 4 m layer of loose sandy soil. The fine-grained content of the soil to be improved was approximately F_c = 5 to 25%, the porosity was n = 39 to 44%, the N-value was 3 to 15, and the liquefaction strength R_{120} was 0.27 to 0.32. The liquefaction strength of the soil to be improved was 0.27 to 0.32. The design earthquake ground motion was assumed to be the level 2 earthquake motion at Airport A. The maximum shear stress ratio based on one-dimensional seismic response analysis for the liquefied area was about

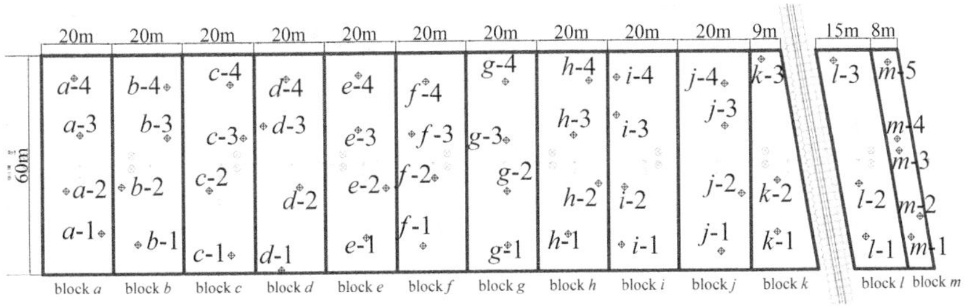

FIGURE 7.4 Range and plan of post-survey points with chemical grouting applied at Site A.

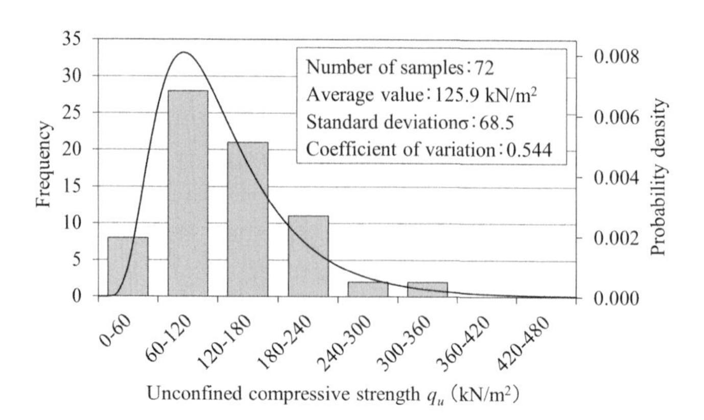

FIGURE 7.5 Distribution of unconfined compression strength in a representative block at Airport A.

0.35, and the design base strength (unconfined compression strength) required for the improved ground by chemical grouting was 60 kN/m^2. The improvement area was the entire 60 m width of the runway, and the improvement depth was GL-4 m to GL-8 m (approximately 4 m improvement thickness). Figure 7.5 shows the frequency distribution of unconfined compression strength in the representative block (Block h) of Airport A.

The mean value is 125.9 kN/m^2 and the coefficient of variation is 0.544 for 72 data samples of unconfined compression strength. Assuming that the distribution of the unconfined compression strength of the block is lognormal, the chi-square value is 3.75, which is smaller than the variable value $C_{(1-\alpha,f)} = 12.6$ at the significance level of 0.05 and 6 degrees of freedom. The distribution of the measured unconfined compression strength in the improved soil by chemical grouting of the block was found to conform to a lognormal distribution. This trend was observed for all blocks except for Block d at Airport A, as well as for the other works shown in the technical manual, and the unconfined compression strength of the ground properly improved by the chemical grouting is generally considered to conform to a lognormal distribution. The authors confirmed by chi-square test that the distribution of measured unconfined compression strength of the improved ground by chemical grouting conforms to a lognormal distribution at a significance level of 5% and then conducted bearing capacity analysis and deformation verification considering strength heterogeneity by Monte Carlo simulation.

On the other hand, Figure 7.6 shows the frequency distribution of unconfined compression strength for the Block d of Airport A.

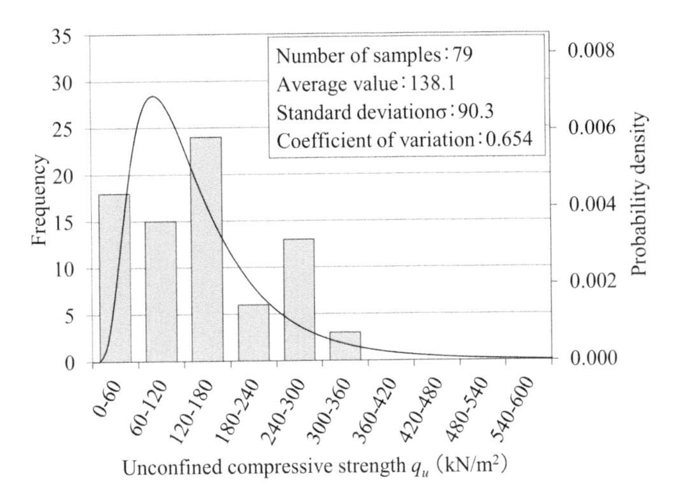

FIGURE 7.6 Distribution of unconfined compression strength in Block d of Airport A.

A goodness-of-fit test between the unconfined compression strength and the lognormal distribution showed a chi-square value of 32.42, which is greater than the significance level of 0.05 and the variable value $C_{(1-\alpha,f)}$ = 12.6 with 6 degrees of freedom, indicating that the distribution of unconfined compression strength for the same block does not fit a lognormal distribution. In this chapter, a section of ground improved by chemical grouting in Block d of Airport A, where the distribution of unconfined compression strength did not conform to a lognormal distribution, is taken as a representative section, and the bearing capacity and deformation analysis methods for such a case are presented. A comparison of the results with those obtained assuming a lognormal distribution is also presented.

7.2.3.3 Analysis Conditions

Here, out of a total of 13 blocks in the construction area, the cross-section of the improved ground by chemical grouting was modeled in detail in Block d, where the distribution of the unconfined compression strength q_u did not conform to a lognormal distribution based on the results of the ground investigation after the ground improvement work at the site. The mesh spacing within the improvement area was divided into 0.2 m × 0.2 m meshes, taking into account that the autocorrelation distance of the unconfined compression strength in the vertical direction was about 0.2 to 0.3 m, which was confirmed in the ground investigation by the authors.

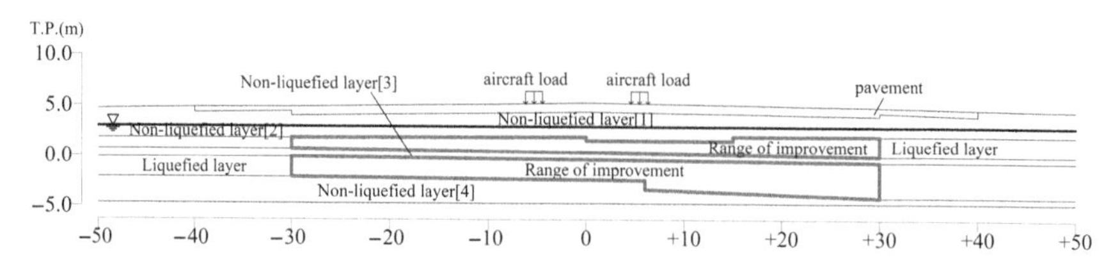

FIGURE 7.7 Analysis model diagram.

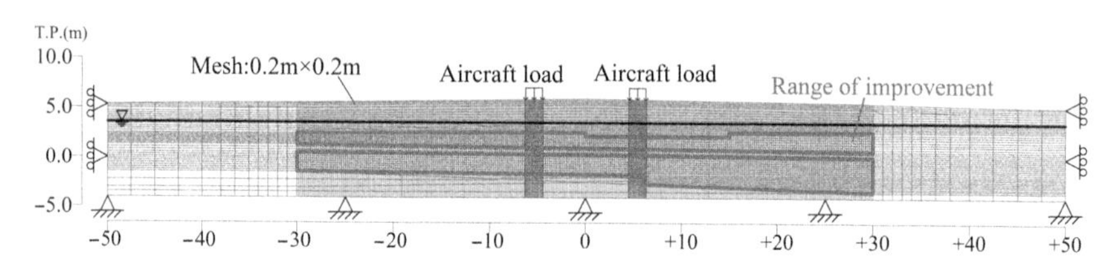

FIGURE 7.8 Analytical mesh diagram.

However, the mesh around the aircraft load was divided more densely to ensure calculation accuracy, with a mesh width of about half that of the mesh around the load. Figure 7.7 shows the analytical model used in the deformation analysis.

Figure 7.8 shows the analytical mesh diagram.

The total width of the pavement is 80 m. The width of the runway in the center of the pavement is 60 m, and the width of the outer shoulder of the runway is 10 m on each side. To simplify the analytical model, the runway and the shoulder were made of the same material. For the boundary conditions, the bottom of the analytical domain was assumed to have fixed displacements in the X and Y directions, and the sides of the analytical domain were assumed to have fixed displacements in the X direction and free displacements in the Y direction.

The bearing capacity analysis performed in this chapter is a calculation method that adjusts the shear resistance of the model as a whole. For this reason, the models used in the bearing capacity analysis do not model pavements that have extremely high shear resistance in the model. However, the self-weight of the pavement was considered by equally distributed loading. The constitutive equation of the improvement area and the original ground in the bearing capacity analysis was assumed to be elasto-perfectly plastic (Mohr-Coulomb model), the failure criterion was Mohr-Coulomb, and the plastic potential was the relevant flow law. The required analytical constants are unit volume weight γ, modulus of elasticity E_{50}, shear strength c, and

Poisson's ratio v. Representative values were used based on the results of soil investigations at Airport A. The shear strength of the non-liquefied layer was determined from the results of triaxial compression tests. The shear strength of the liquefied layer was set to zero, assuming full liquefaction.

In order to evaluate the shear strength c in the improvement area by chemical grouting, the spatial distribution of the unconfined compression strength of each element was calculated considering the spatial heterogeneity using the random field theory. The constitutive equations for each material in the deformation analysis were set to linear elasticity for the pavement and nonliquefied layer, and to an exponential function model for the improvement area and liquefied layer, which is the mechanical model used in FLIPDIS for consolidation deformation analysis. The parameters a and b of the exponential model were set to the general values ($a = 9$, $b = 4$) given by Morio et al. Other required analytical constants are unit volume weight γ, Poisson's ratio v, elastic modulus E_0, and hydraulic conductivity k. Table 7.2 lists the analytical constants set for bearing capacity analysis.

The posterior results for the Block d analyzed showed that the unconfined compression strength compliance was 78%, the average unconfined compression strength was 138 kN/m^2, and the coefficient of variation was 0.65.

Figure 7.9 shows the wheel arrangement of the B777-9 and the loads used in the analysis.

Table 7.4 shows the results of the aircraft load settings.

Since this bearing capacity analysis was performed in the transverse direction of the runway using plane strain elements with a depth width of 1 m, the loads (P_1 and P_2) were considered in the direction of the analysis section in Figure 7.9. In the depth direction of the analytical cross-section, the distributed load was set considering the load dispersion up to the top of the roadbed (GL-1 m), because the load is dispersed in the roadbed. In the direction of the analytical cross-section, on the other hand, load dispersion is taken into account in the analysis, and therefore the range of load dispersion in the ground is not considered when setting the distributed loads. In addition, a 30% increase in the static load was taken into account as the impact load due to the running of the aircraft. The aircraft load is the maximum load that can be assumed, so variations in load are not considered.

TABLE 7.2 List of Analytical Constants Set in Bearing Capacity Analysis

Layer	Soil	Unit Weight		N-Value	Modulus of Elasticity	Shear Strength		Poisson's
		Wet γ_t (kN/m³)	Saturated γ_{sat} (kN/m³)		E_{50} (kN/m²)	c (kN/m²)	φ (°)	Ratio ν
Non-liquefied layer [1]	Sandy	18.9	19.4	11	30,500	8.3	37.7	0.33
Non-liquefied layer [2]	Clay	16.5	16.6	3	9500	18.5	18.4	0.33
Improvement area	Sandy	18.2	19.0	—	23,100	[a]	0.0	0.33
Liquefied layer	Sandy	18.2	19.0	8	21,000	0.0	0.0	0.33
Non-liquefied layer [3]	Clay	16.5	16.6	1	3600	18.5	18.4	0.33
Non-liquefied layer [4]	Sandy	17.8	17.8	20	57,100	13.7	35.8	0.33

[a] Based on the distribution of measured unconfined compression strength based on the results of ex-post investigation.

TABLE 7.3 List of Analytical Constants Set in Deformation Analysis

Layer	Soil	Unit Weight		N-Value	Poisson's Ratio v	Modulus of Elasticity E_0 (kN/m^2)	Coefficient of Permeability k (cm/s)	Parameter	
		Wet γ_i (kN/m^3)	Saturated γ_{sat} (kN/m^3)					a	b
Pavement	—	22.5	22.5	—	0.35	414,000	1.0E-07	—	—
Non-liquefied layer [1]	Sandy	18.9	19.4	11	0.33	194,000	9.5E-03	—	—
Non-liquefied layer [2]	Clay	16.5	16.6	3	0.33	88,000	7.1E-05	—	—
Improvement area	Sandy	18.2	19.0	—	0.33	[a]	1.2E-06	9	4
Liquefied layer	Sandy	18.2	19.0	8	0.33	122,000	3.1E-03	9	4
Non-liquefied layer [3]	Clay	16.5	16.6	1	0.33	82,000	1.0E-06	—	—
Non-liquefied layer [4]	Sandy	17.8	17.8	20	0.33	205,000	2.3E-02	—	—

[a] $E_0 = 122{,}000 + 12{,}200 \times q_u/q_{uck}$ (unconfined compression strength q_u is set to the value of each element based on random field theory).

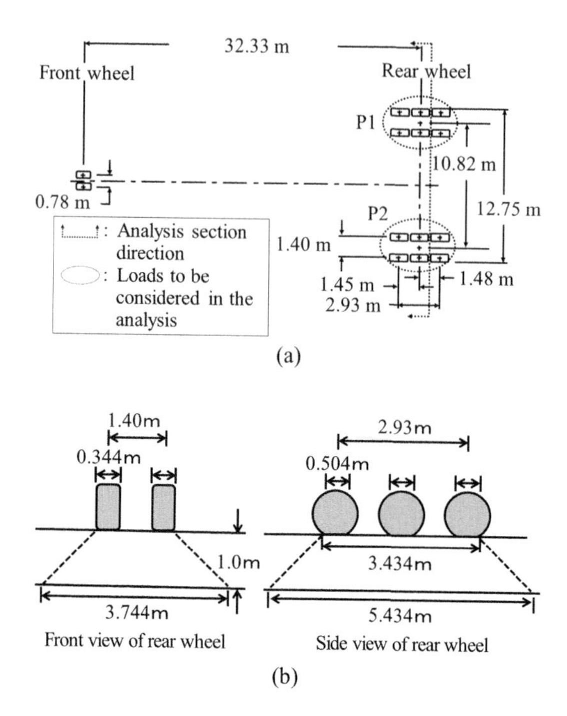

FIGURE 7.9 Wheel arrangement of B777-9 aircraft and loads considered in the analysis (a) wheel layout, (b) load sharing range.

TABLE 7.4 Aircraft Load Setup Results

Aircraft		**B777-9**
Ground contact pressure		1.58 (N/mm²)
Ground contact width		344 (mm)
Ground contact length		500 (mm)
Loads per a wheel P_0		$P_0 = 1.58 \times 344 \times 500 \div 1000 ≒ 272$ (kN)
Aircraft loads considering an impact load with 30% of static load	P_1	$P_1 = 6 \times P_0/(1.400 + 0.344)/(2.930 + 0.504 + 1.000 \times 2) \times 1.3 ≒ 224$ (kN/m²) *Loading width on analysis model: 1.744 m
	P_2	$= P_1$

7.2.3.4 Bearing Capacity Analysis Results

Figure 7.10 shows an example of the maximum shear strain distribution in Block d.

Figure 7.11 shows the relationship between the degree of safety of bearing capacity and the number of iterations in Block d.

Figure 7.12 shows the frequency distribution of the degree of safety of bearing capacity.

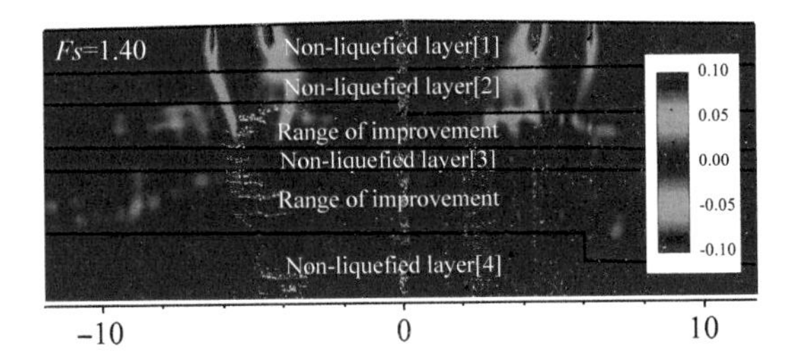

FIGURE 7.10 Example of maximum shear strain distribution in Block d.

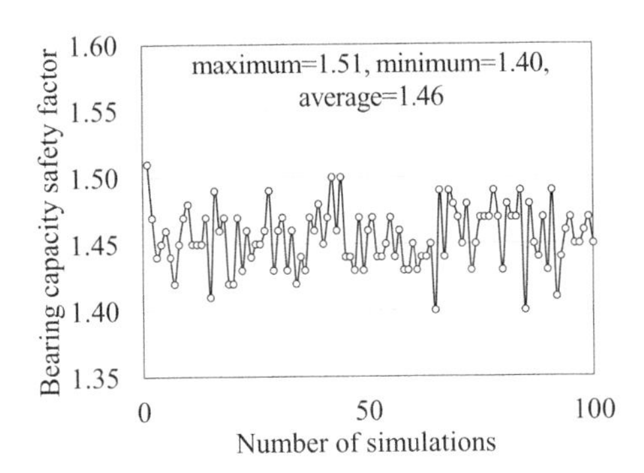

FIGURE 7.11 Bearing capacity safety factor of Block d and number of iterations.

The bearing capacity safety factor of the Block d varied between 1.40 and 1.51, with an average value of 1.46. Figure 7.12 also shows the results for the case assuming a lognormal distribution. The results show that the bearing capacity safety values for the cases assuming a lognormal distribution varied between 1.45 and 1.55, with an average value of 1.50. The maximum shear strain is greater than 1.0 in all cases, which indicates that the bearing capacity for post-seismic aircraft loads is satisfactory, although the maximum shear strain extends into the improvement range.

7.2.3.5 Deformation Analysis Results

Figure 7.13 shows the distribution of ground surface subsidence occurring at the runway and shoulder.

Figure 7.13 shows the distribution of cases in which the maximum or minimum subsidence of the ground surface immediately after the earthquake, three days after the earthquake, and after the dissipation of excess pore water pressure—which were obtained from 100 Monte Carlo simulations—is located near the runway center (x = −20 m to +20 m section). Figure 7.14 shows the distribution of the amount of change in gradient of the ground surface that occurs in the center of the runway (x = −20 m to +20 m section).

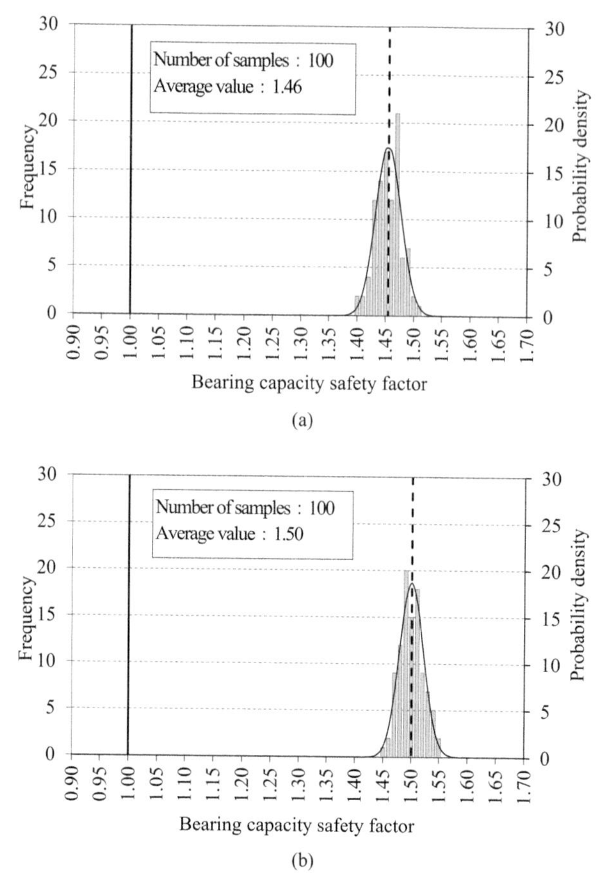

FIGURE 7.12 Example of frequency distribution of bearing capacity safety of Block d, (a) cases based on measured distributions, (b) cases assuming lognormal distributions.

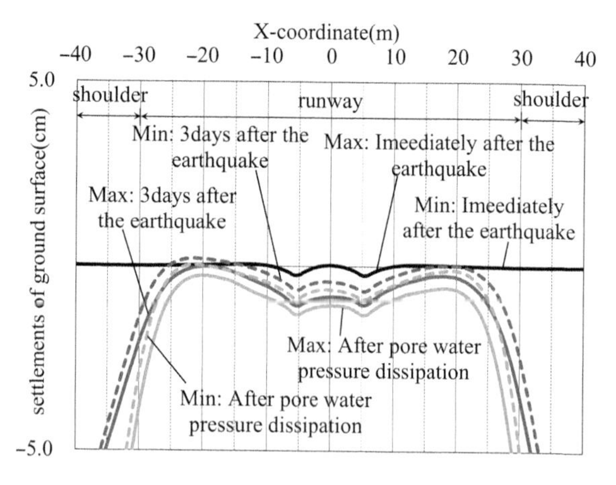

FIGURE 7.13 Distribution of ground surface settlement of Block d.

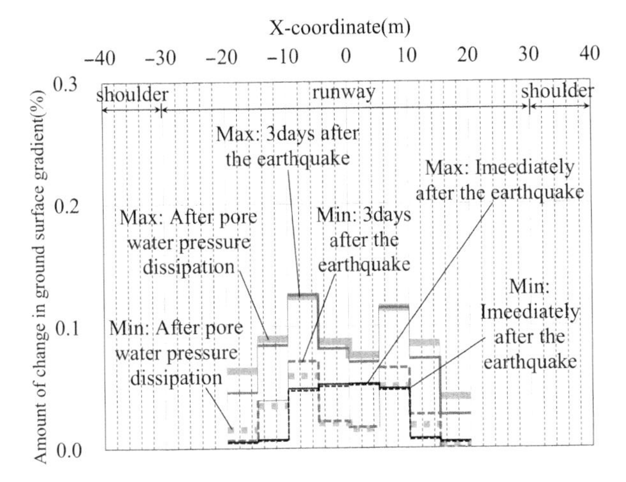

FIGURE 7.14 Distribution of change in ground surface gradient for Block d.

Figure 7.14 shows the distribution of cases in which the change in gradient immediately after the earthquake, three days after the earthquake, and after the dissipation of excess pore pressure is the largest or smallest near the runway center ($x = -20$ m to +20 m section), as obtained from 100 Monte Carlo simulations. The denominator for calculating the gradient was set to 5 m, which is the measurement interval in the crosswise direction during the maintenance of Airport A. Figure 7.15 shows the frequency distribution of the maximum gradient change of the ground surface that occurs in the center of the runway.

Figure 7.15 also shows the results for the case assuming a lognormal distribution. The results show that the maximum gradient change is within the design limit (0.2%) in all phases for all cases.

7.2.3.6 Performance Evaluation Results

The authors assumed that the unconfined compression strength of the ground improved by chemical grouting at Airport A conforms to a lognormal distribution, expressed the shear strength using random field theory, and performed bearing capacity and deformation analyses considering strength heterogeneity by Monte Carlo simulation using the finite element method and the shear strength reduction method. Furthermore, the authors proposed a chart diagram that enables bearing capacity verification based on the performance specifications once the unconfined compression strength and the coefficient of variation of the improved soil due to chemical grouting are obtained. Figure 7.16 plots the results of bearing

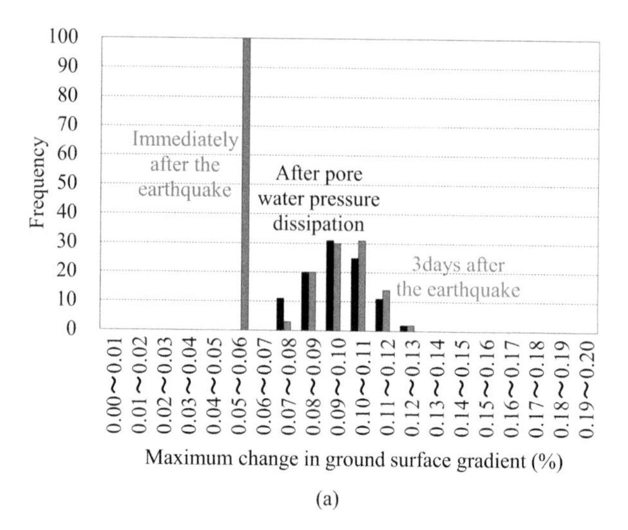

(a)

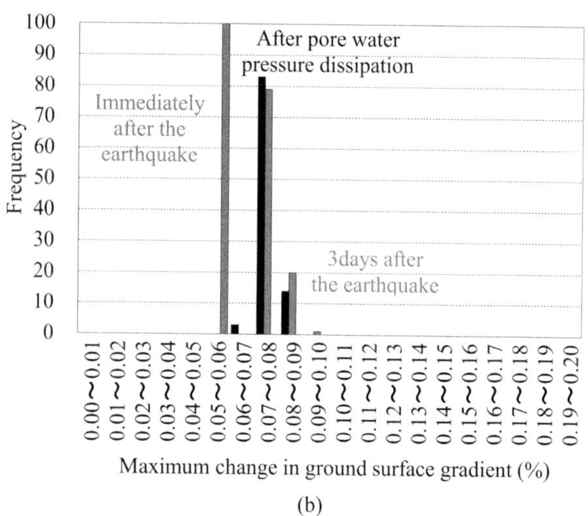

(b)

FIGURE 7.15 Distribution of maximum change in ground surface gradient for Block d, (a) cases based on measured distribution, (b) cases assuming lognormal distributions.

capacity and deformation analysis for the Block d of Airport A modeled in detail on the "relationship between bearing capacity safety and fitness factor" shown by the authors.

Figure 7.16 also shows the results for the case assuming a lognormal distribution. The results are plotted in a range that satisfies the bearing capacity and slope performance proposed by the authors and are consistent with our chart.

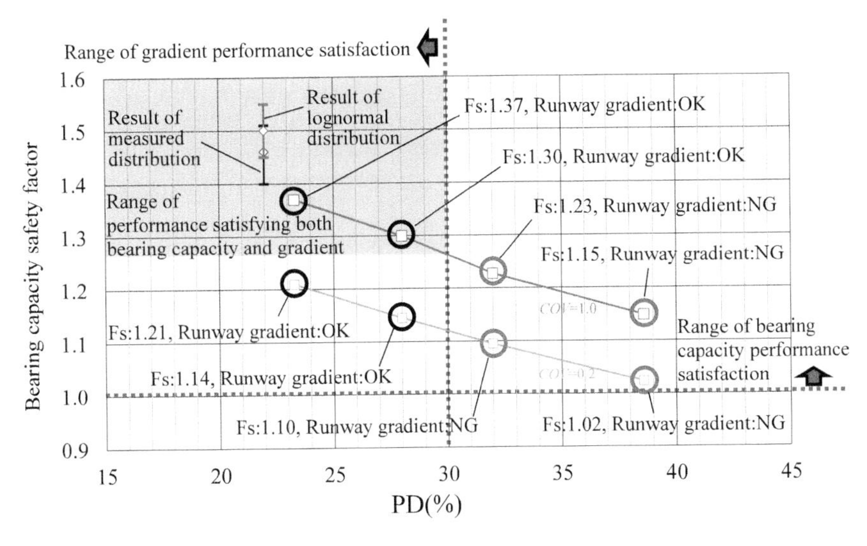

FIGURE 7.16 Comparison with the chart for checking the precision ratio against the bearing capacity safety and deformation.

7.2.4 What to Do If NG Is Found in the Primary Evaluation

7.2.4.1 Analysis Conditions

Of the total 13 blocks in the construction area, the cross-section of the improved ground by chemical grouting in Block b—which was NG in the primary evaluation based on the results of the ground investigation after the ground improvement work at the site—was modeled in detail. Figure 7.17 shows the analytical model used in the deformation analysis.

Figure 7.18 shows the analytical mesh diagram.

Figure 7.19 shows the frequency distribution of unconfined compression strength in Block b.

The average value of unconfined compression strength was 127.3 kN/m^2 with a coefficient of variation of 0.774 for the 80 data samples. Assuming that the distribution of unconfined compression strength of the block is lognormal, the chi-square value is 9.54, which is smaller than the variable

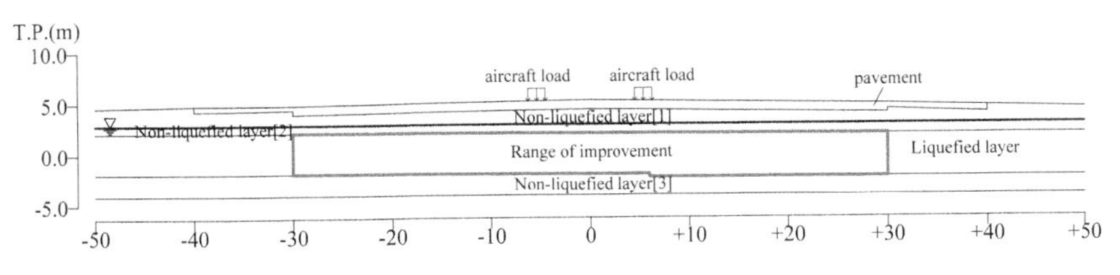

FIGURE 7.17 Analytical cross-section (Block b).

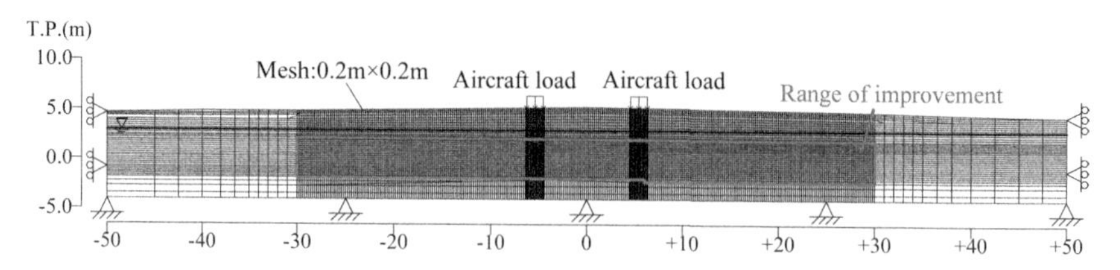

FIGURE 7.18 Analytical mesh diagram (Block b).

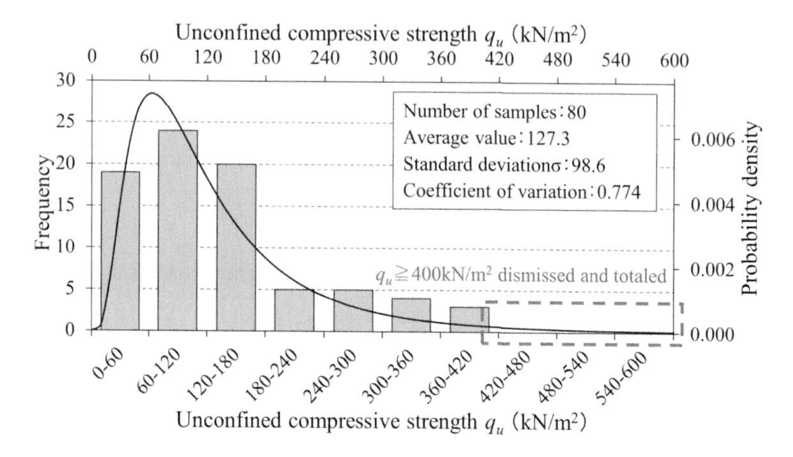

FIGURE 7.19 Frequency distribution of unconfined compression strength.

value $C_{(1-\alpha,f)} = 12.6$ at a significance level of 0.05 with 6 degrees of freedom. The distribution of the measured unconfined compression strength in the improved soil by chemical grouting of the block was found to conform to a lognormal distribution.

Other analysis conditions are basically the same as those described in "Section 7.2.3.3 Analysis Conditions," so please refer to that section.

7.2.4.2 Bearing Capacity Analysis Results

Figure 7.20 shows an example of the maximum shear strain distribution in Block b.

Figure 7.21 shows the relationship between the degree of safety of bearing capacity and the number of iterations in Block b.

Figure 7.22 shows the frequency distribution of the degree of safety of bearing capacity.

The bearing capacity safety factor of Block b varied between 1.34 and 1.47, with an average value of 1.42. In all cases, the values were greater than 1.0, indicating that the bearing capacity for post-seismic aircraft

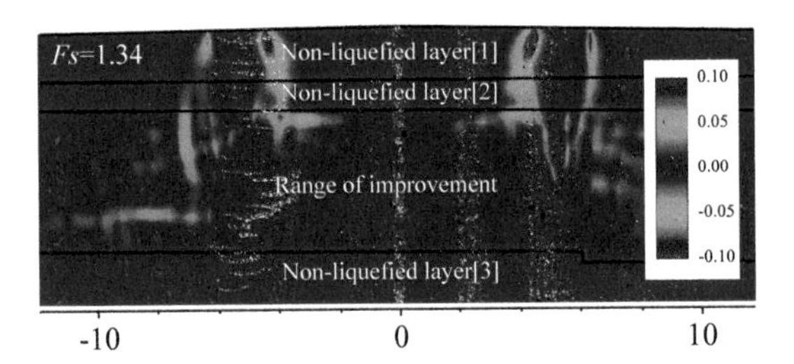

FIGURE 7.20 Maximum shear strain distribution (Block b).

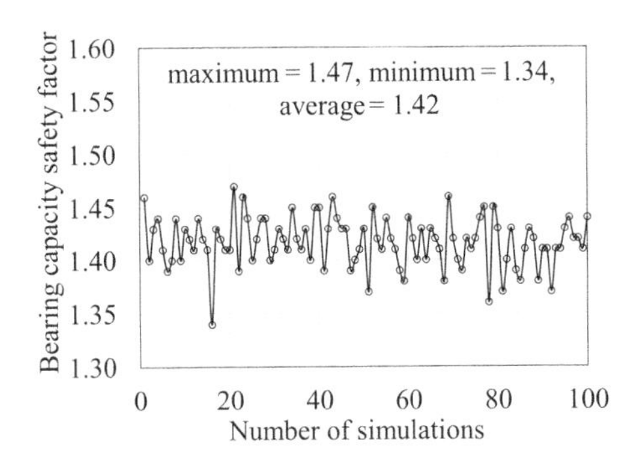

FIGURE 7.21 Bearing capacity action ratio and number of calculations (Block b).

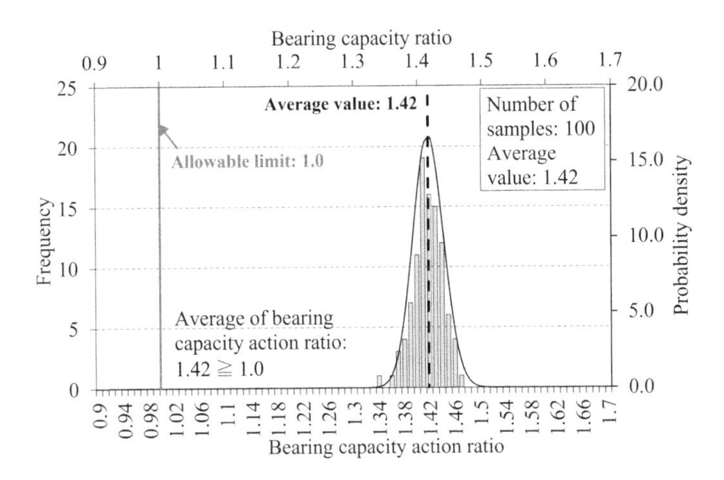

FIGURE 7.22 Histogram of bearing capacity action ratio of bearing capacity (Block b).

loads is satisfactory, although the maximum shear strain extends into the range of the modification.

7.2.4.3 Deformation Analysis Results

Figure 7.23 shows the distribution of surface subsidence occurring on the runway and shoulders.

Figure 7.23 shows the distribution of cases in which the maximum or minimum subsidence occurs near the center of the runway ($x = -20$ m to $+20$ m) immediately after the earthquake, three days after the earthquake, and after the dissipation of excess pore pressure, as obtained from 100 Monte Carlo simulations.

Figure 7.24 shows the distribution of the amount of gradient change of the ground surface that occurs at the center of the runway ($x = -20$ m to $+20$ m section).

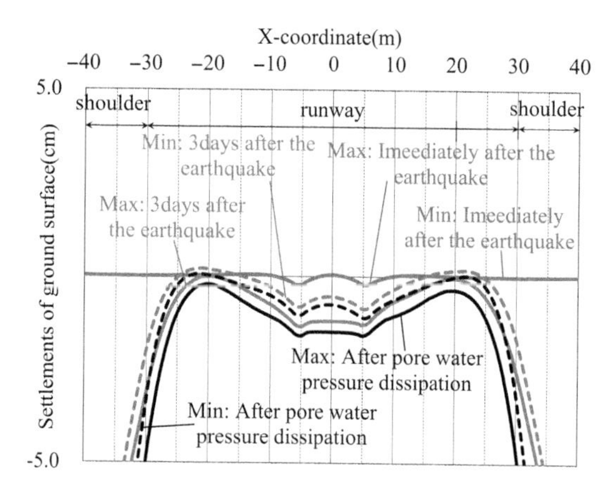

FIGURE 7.23 Distribution of ground surface settlement of Block b.

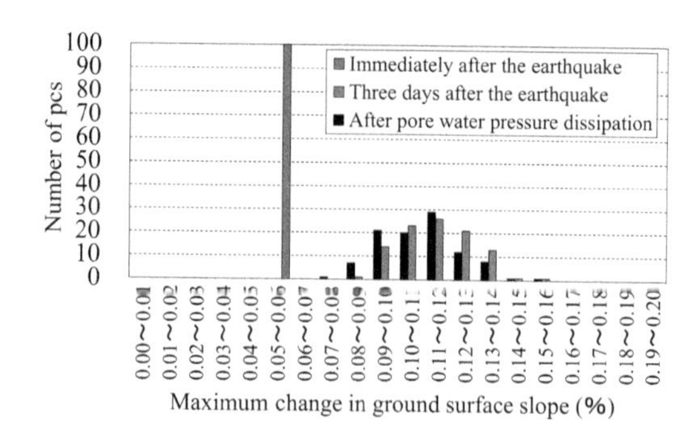

FIGURE 7.24 Distribution of change in ground surface gradient for Block b.

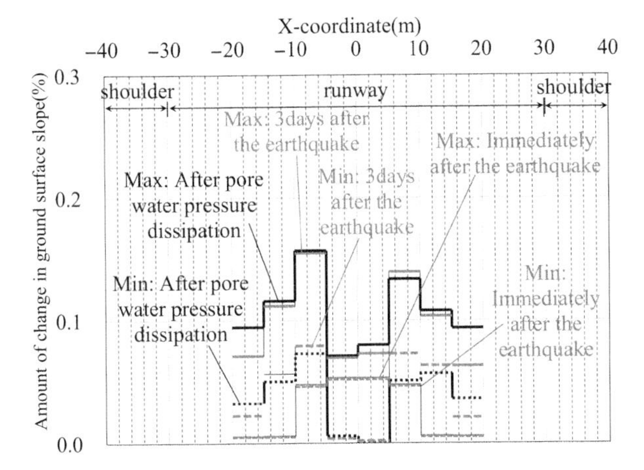

FIGURE 7.25 Distribution of maximum change in ground surface gradient for Block b.

Figure 7.24 shows the distribution of the cases in which the maximum or minimum change in gradient occurs near the runway center (x = −20 m to +20 m) immediately after the earthquake, three days after the earthquake, and after the dissipation of excess pore pressure, as obtained from 100 Monte Carlo simulations. The denominator for calculating the gradient was set to 5 m, which is the measurement interval in the crosswise direction during the maintenance of Airport A. Figure 7.25 shows the frequency distribution of the maximum gradient change of the ground surface that occurs in the center of the runway.

Figure 7.25 also shows the results for the case, assuming a lognormal distribution. The results show that the maximum gradient change is within the design limit (0.2%) in all phases of all cases.

7.3 CONCLUSIONS

In this chapter, based on the verification in Chapters 5 and 6, a new performance evaluation method based on performance specifications was proposed for the bearing capacity and flatness required of airport runways after earthquakes. The proposed performance evaluation method was also applied to a case study of the actual ground surface at Airport A, which was liquefied by the chemical grouting. As a result, the following were found:

1) The authors proposed a chart diagram that enables bearing capacity and deformation verification based on the performance specifications, based on the unconfined compression strength of the

previously consolidated soil and its variation, by evaluating the relationship between the bearing capacity safety and the precision ratio—together with the results of slope verification of the runway center and the precision ratio.

2) Based on the chart confirming the precision ratio for the proposed bearing capacity safety, the performance evaluation for Airport A was conducted. As a result, it was confirmed that the bearing capacity and deformation were satisfied for all blocks except for Blocks b and e.

3) The results of the goodness-of-fit tests of the unconfined compression strength and the lognormal distribution for Block d at Airport A showed that the block did not locally fit the lognormal distribution, but the detailed modeling of the ground conditions and the bearing capacity and deformation tests showed that the block satisfied both the bearing capacity safety factor for airport runways (1.0) and the allowable gradient change (0.2%). The results also satisfied the performance in the case where a lognormal distribution was assumed.

4) The results of the bearing capacity and deformation analysis for the Block d of Airport A modeled in detail in this chapter are plotted against the "Relationship between bearing capacity safety and precision ratio" shown in Chapter 5. The results were plotted within the range satisfying the bearing capacity and gradient performance proposed by the authors, confirming the applicability of this chart diagram.

5) A section of improved ground by chemical grouting in Block b, which was NG in the primary evaluation, was taken as a representative section, and the bearing capacity and deformation analysis for such a case were presented. Comparison of the results with the results assuming a lognormal distribution was also presented, and it was confirmed that the bearing capacity and deformation were satisfactory.

REFERENCE

Davison, A.C.: *Bootstrap methods and their application*, Cambridge University Press, 1997.

CHAPTER 8

Quality Control and Assurance for Improved Ground by Chemical Grouting

Teruhisa Fujii, Tomoyuki Kaneko, and Yasutaka Kimura

8.1 INTRODUCTION

The following is a method for investigating the effect of ground improvement on ground improved by chemical grouting (hereinafter referred to as "improved ground"), as indicated in the administrative communication by the Ministry of Land, Infrastructure, Transport and Tourism (hereinafter referred to as "the administrative communication") (MLIT, 2017).

The field investigation items include sampling by boring and dynamic cone penetration tests. As shown in Figure 8.1 (MLIT, 2017), the survey point is located at 1/2 of the radius of the improvement, with a separation of about 50 cm to ensure there is no disturbance in the sampling process.

The number of borehole investigations is about three for soil volumes of less than 5000 m^3 and one additional borehole for each additional 2500 m^3 of soil volume of 5000 m^3 or more to be improved. The standard depth of investigation by the unconfined compression test and the dynamic cone penetration test using collected samples is from the top to the bottom of the improvement area.

DOI: 10.1201/9781032670133-8

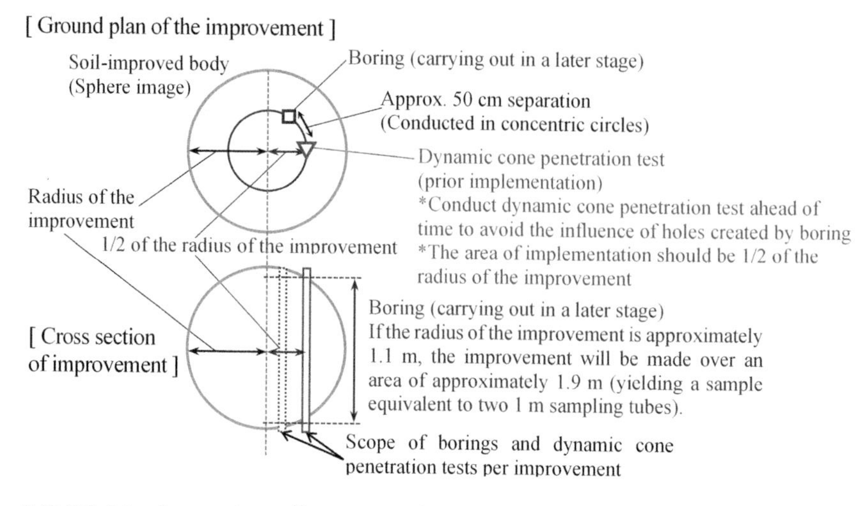

FIGURE 8.1 Separation of borings and soundings.

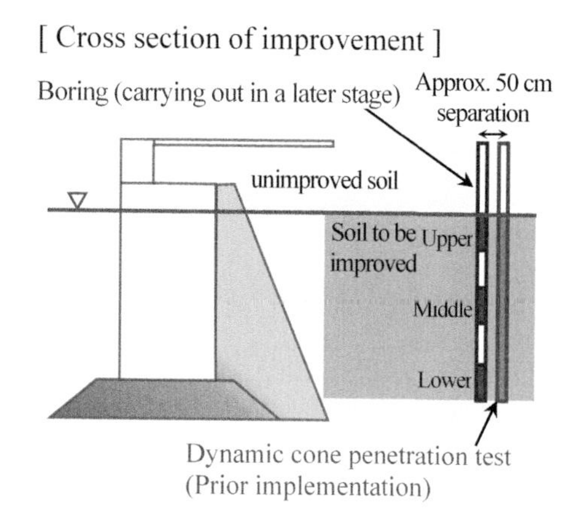

FIGURE 8.2 Survey depth.

When the thickness of the improvement layer is 6 m or thicker, three depths are selected per borehole: "Upper," "Middle," and "Lower," as shown in Figure 8.2 (MLIT, 2017).

Table 8.1 (CDIT, 2020) shows the actual results of the amount of soil to be improved and the number of specimens in the posterior investigation.

The aforementioned quality control method is applied to improved ground that has been improved so that the implemented injection rate, defined as the ratio of the implemented injection rate to the design injection rate, is 100% in a ground improvement project using chemical grouting.

The number of unconfined compression strength q_u in the depth direction that can be obtained is reduced when the thickness of the improvement

TABLE 8.1 Amount of Improved Soil and Number of Specimens in Posterior Study

Name of Construction	Amount of Improved Soil (m^3)	Number of Samplings	Quantity of Unconfined Compression Test (Sample)	Sampling Point
Construction of Runway C at Tokyo International Airport	42,000	14	40	Upper, middle, lower
New Chitose Airport Runway A north side construction	6900	4	10	Upper, middle, lower
Misaki Fishing Port Improvement Project	3200	4	7	Upper, lower
Oe Pier Improvement Work	16,800	6	18	Upper, middle, lower
Fukui Port Coastal Construction	8600	10	16	Upper, middle, lower

Note: One sample consists of three specimens, and the number of samples is selected according to the layer thickness of the target soil.

layer is thin, so the survey is not designed to confirm the overall quality in the ground, but to guarantee a minimum quality by spot checks.

Therefore, there is a concern that the accuracy of statistical processing to evaluate the mean value and variation of the unconfined compression strength q_u may be reduced for improved soils with heterogeneity due to an implementation injection rate of less than 100%. This means that the number of investigations necessary to evaluate the quality and performance of the improved ground must be ensured, especially when the injection rate is less than 100%.

Therefore, this chapter proposes a method for setting the appropriate number of investigations required to evaluate the strength of improved ground in assessing the effect of improvement of spatially heterogeneous improved ground, based on the content of the new strength evaluation method at Airport A shown in Chapter 4 as an example of improved ground investigation.

8.2 CONDITIONS FOR SETTING THE NUMBER OF SURVEYS

At Airport A, as shown in Figure 8.3 (CDIT, 2020), the improvement specifications are divided into a total of 13 blocks (blocks a through m) in plan for design and construction purposes.

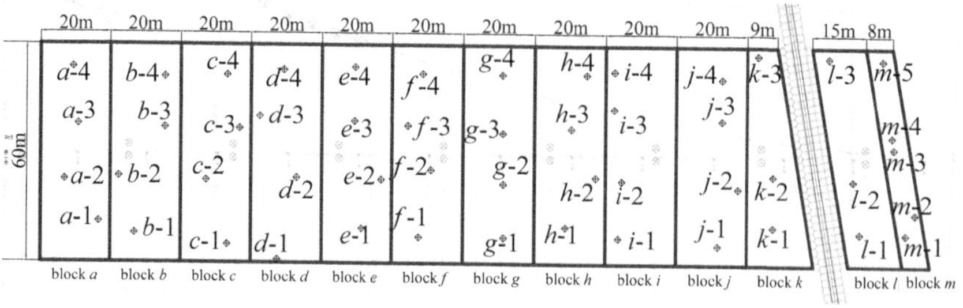

FIGURE 8.3 Range and plan of post-survey points with chemical grouting applied at Airport A.

The "0 points" shown hereafter indicate the number of survey points arranged in planar form, and the "number of surveys" indicates the number of unconfined compression strengths to be obtained in a given improvement volume. The "0 depth" indicates the number of surveys in the direction of depth to be acquired per site.

In Airport A, as mentioned above, post-construction investigations were conducted at all 51 points shown in Figure 8.3, and the calculated and measured values of 897 unconfined compression strengths q_u and estimated unconfined compression strengths q_u were obtained by in-situ and laboratory tests, as shown in Figure 8.4 (Kasama et al., 2022).

The mean unconfined compression strength q_u and estimated unconfined compression strength q_u were 137.1 kN/m^2, standard deviation 85.5, and coefficient of variation 0.623. It was confirmed that the distribution of

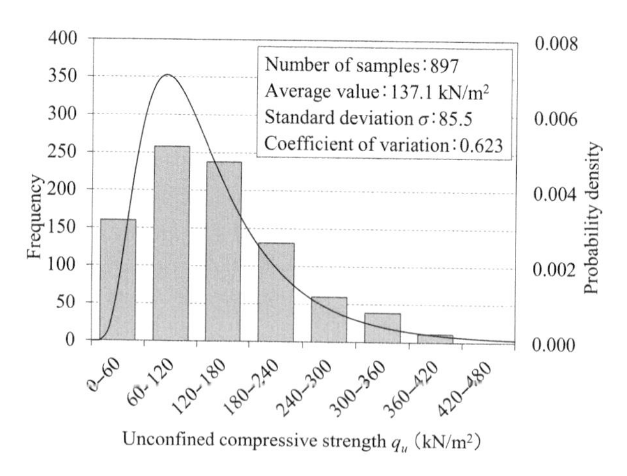

FIGURE 8.4 Distribution of q_u and estimated q_u at Airport A.

unconfined compression strength q_u and estimated unconfined compression strength q_u fit a log normal distribution. The autocorrelation distances were about 0.2 to 0.3 m in the vertical direction and 2 m in the horizontal direction (Kasama et al., 2022) for the improved soil, while they were about 1 to 5 m in the vertical direction and 15 m in the horizontal direction (Matsui, 1992) for the naturally deposited soil.

The autocorrelation distance of the improved ground is about one-tenth that of the naturally deposited ground, indicating that the autocorrelation distance of the improved ground is shorter than that of the naturally deposited ground and that the improved ground can be evaluated as random. In accordance with the current administrative Communication (CDIT, 2020), the number of survey points per block is 3, and the depth per point is 2, thus the total number of survey points is 2*3 = 6. When this is applied to Airport A, there are a total of 13 blocks from a to m, so the total number of surveys for the entire construction area is about 80.

In the construction management of the chemical grouting, the injection work is performed to achieve a 100% implementation injection rate, which is defined as the ratio of the implementation injection rate to the design injection rate. However, there are cases in which the injection rate does not reach 100%, and in these cases the unimproved ground is not improved to the extent of the improved amount, resulting in a heterogeneous state in which the improved and unimproved portions are mixed together.

In evaluating the strength of such improved ground with a mixture of improved and unimproved sections, a larger number of investigations is required. Therefore, in this chapter, the number of surveys is proposed according to the implemented injection rate obtained after construction.

One of the indicators in the quality evaluation of improved ground is the conformance ratio (Kasama et al., 2022). The compliance ratio is the percentage of volume in which the unconfined compression strength q_u of the improved soil satisfies the design basis strength. Figure 8.6 and Table 8.2 show an example of the relationship between the compliance rate and the implemented injection rate for the improved ground at Airport A shown in Figure 8.5.

In Figure 8.6, the compliance ratio is the value obtained by calculating the proportion of the unconfined compression strength q_u for each block that satisfies the design basis strength for the population, using the probability distribution obtained by assuming a lognormal distribution for the unconfined compression strength q_u. The implementation injection rate is

TABLE 8.2 Relationship Between Precision Ratio and Rate of Injection Implemented (Case Study of Airport A)

Block	Rate of Injection Implemented (%)	Precision Ratio (%)
a	63.5	91
b	61.7	77
c	58.5	94
d	46.2	86
e	70.7	62
f	69.4	83
g	60.1	94
h	69.6	89
i	66.2	80
j	70.6	95
k	74.2	92
l	81.5	97
m	90.4	83

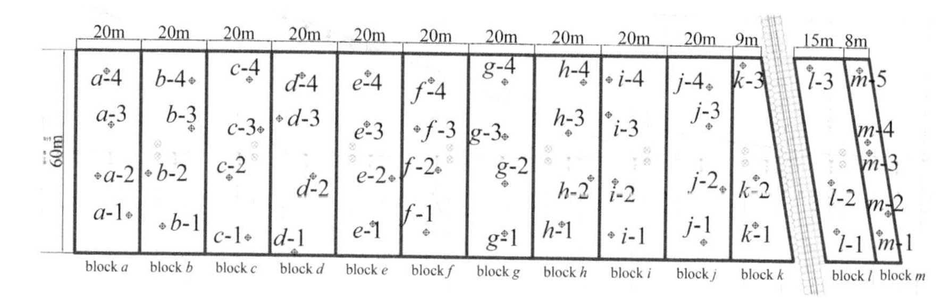

FIGURE 8.5 Range and plan of post-survey points with chemical grouting applied at Airport A.

a value calculated as the ratio of the implemented injection volume to the design injection volume for each block.

The reasons why the implemented injection rate was less than 100% include the fact that, in general, the chemical grouting rate is often set on the safe side in the design stage to avoid insufficient chemical grouting in response to uncertainties in the formation composition and porosity of the target soil, and that the chemical grouting rate may have been limited by the relatively high fine grain content mixed in the soil to be improved.

As shown in Figure 8.6, the conformance rate tends to exceed the implementation injection rate. This is because the chemical grouting volume is

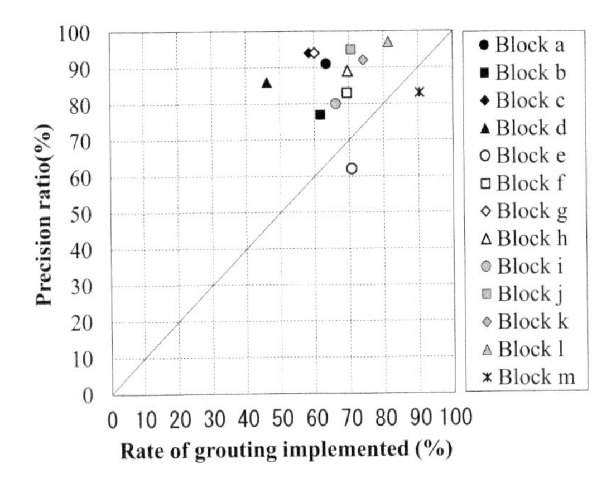

FIGURE 8.6 Relationship between precision ratio and percentage of injections implemented rate (case study of Airport A).

set on the safe side, as described above. In this chapter, we propose the number of surveys relative to the rate of implementation injection, assuming that the rate of compliance and the rate of implementation injection are comparable, since the compliance rate is usually not yet identified in the planning phase of post-construction surveys.

In addition, the *Technical Manual for Permeation Grouting Method* (CDIT, 2020) states that the distribution of unconfined compression strength q_u of the improved soil conforms to the standard normal distribution. In general, it is known that if the number of data is 30 or more, the distribution is roughly equivalent to the standard normal distribution (UTokyo, 1991). Therefore, the number of investigations in this chapter was set based on the belief that at least 30 datasets are necessary to confirm the strength of the improved ground.

In this chapter, the survey location of the improved ground is set according to the heterogeneity of the ground in question. In such cases, the survey location should be set after confirming the autocorrelation of the improvement strength of the improved ground. The improved ground at Airport A has been confirmed to have almost no autocorrelation, i.e., it is random (Kasama et al., 2022).

In addition, the shape of the improved soil is not uniform and the unconfined compression strength, q_u, varies even in improved soils with relatively homogeneous sandy soil (Sugano et al., 2020). In previous examples with chemical grouting, it has been confirmed that the shape of the improved soil is not uniform and the unconfined compression strength, q_u, varies even in improved soils with relatively homogeneous sandy soil.

For such random improved ground, the survey points can be randomly placed as long as a specified number of survey points can be secured within the area of interest. For this reason, the post-survey locations in Airport A were mechanically placed at equal intervals within the target area without considering construction information such as the location of injection holes and the amount of chemical grouting.

8.3 RELATIONSHIP BETWEEN THE IMPLEMENTATION INJECTION RATE AND THE NUMBER OF SURVEYS

In general, in ground improvement design, the chemical grouting rate is often set on the safe side in order to avoid insufficient chemical grouting rate due to uncertainty of the target soil. Therefore, for example, if the layer to be improved contains clay, which is not taken into account in the design of ground improvement, it is possible that the improved ground is constructed without any unimproved points even if the implemented injection rate is less than 100%.

On the other hand, if the porosity of the target soil is as assumed in the ground improvement design and the amount of chemical that can actually be improved matches the design value, a large number of unimproved areas will be mixed in as the implementation injection rate is lowered. In evaluating such improved soils, an appropriate number of investigations is required to ascertain the variation in unconfined compression strength q_u and compliance ratio.

Based on the above, this chapter assumes that the compliance rate and the implementation injection rate are similar and proposes the number of surveys required according to the implementation injection rate that is known before the posterior survey. Equation (8.1) (Ang and Tang, 2007) shows the relationship between the compliance rate (implementation injection rate), the number of surveys, and the error that occurs in the obtained compliance rate. Assuming that the conformance rate here is equivalent to the rate of injection, it is possible to set the number of surveys based on the rate of injection known at the planning stage of the post-construction survey.

$$Z = k_{(1-\alpha/2)} \sqrt{\frac{\hat{p}\left(1-\hat{p}\right)}{n}} \tag{8.1}$$

where,

Z: Error

$k_{(1-\alpha/2)}$: Standard normal random variable (1.96 at 95% confidence level)

$1-\alpha/2$: Cumulative probability ($\alpha/2$=0.025 at 95% confidence level)

$\hat{p}$: Conformance rate (percentage of data where unconfined compression strength q_u satisfies the design basis strength, assumed to be consistent with the implementation injection rate as described above)

$1-\hat{p}$: Percentage of non-conformity (percentage of unconfined compression strength q_u not satisfying the design basis strength)

n: Number of surveys.

Figure 8.7 shows the number of surveys and the theoretical line of error for each fit rate from 60 to 98% based on the relationship in Equation (8.1) (Ang and Tang, 2007).

The plots in Figure 8.7 show the actual relationships observed in the post-construction survey at Airport A.

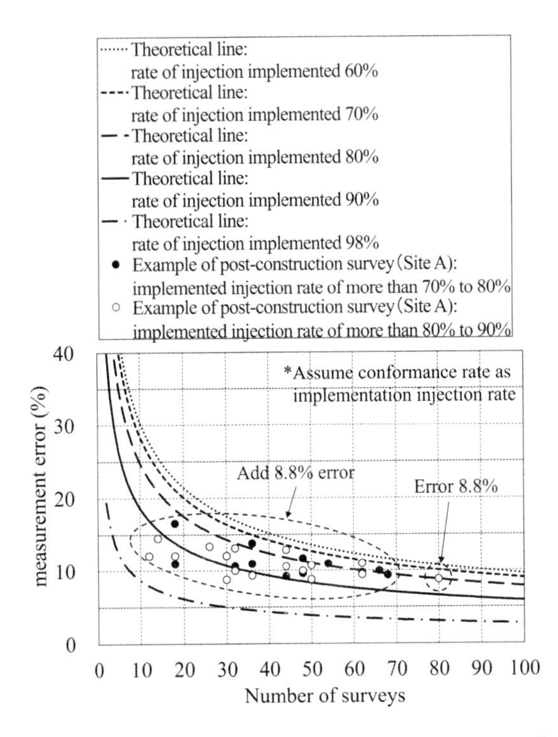

FIGURE 8.7 Relationship between the number of surveys and the margin of error.

On one block at Airport A, 80 data were obtained for the unconfined compression strength q_u of the improved soil at a total of five locations. The total number of surveys was 80 data for these five sites, and the compliance rate was calculated based on the total number of surveys. Next, the number of target sites was reduced from four to one for all combinations within the five sites, and the percent compliance was calculated for each number of surveys.

Since the total number of surveys 80 is the largest number of surveys in the target block, the error in the total number of surveys 80 is considered to be the smallest compared to the errors in the other surveys. The error of the fit rate for all 80 surveys is assumed to be zero, and the difference between the fit rate for each survey is calculated and used as the error for each survey.

The results of the analysis showed that, for a total number of 80 surveys, there were 64 surveys with unconfined compression strength q_u satisfying the design basis strength, and the conformity rate was 80%. The difference between this fit rate of 80% and the corresponding fit rate for each number of surveys was obtained in the range of 0 to 8%. However, it is necessary to consider that the 80% fit rate calculated using the total number of surveys of 80 also includes an error.

Therefore, based on the theoretical line shown in Figure 8.7, the error of 8.8% corresponding to 80 surveys and 80% conformance rate is defined as the error of the conformance rate for the total number of surveys of 80. The difference between the total number of surveys 80 and the percentage of fit for each survey, plus the error of 8.8% mentioned above, is plotted in Figure 8.7 as the error in the percentage of fit for each survey.

In this regard, in this chapter, fit rates for each number of surveys were obtained in the range of 70 to 90%, with all values around 80%. The error on the theoretical line corresponding to 80 surveys is 8.8% for an 80% compliance rate, while the error corresponding to 70 to 90% compliance rates is 10.0 to 6.6%, all within a very small range of 1–2%.

Therefore, it was deemed reasonable to uniformly add the 8.8% error corresponding to the number of surveys 80 to the errors for the other surveys. The error for each number of surveys ranged from 9 to 17%, indicating that the theoretical line tended to be generally consistent with the survey cases in Airport A.

Based on the above, it was evaluated that this theoretical line is applicable to the survey data of improved ground, and the proposed number of post-construction surveys of improved ground is based on the theoretical line.

8.4 NUMBER OF INVESTIGATIONS REQUIRED TO CONFIRM THE STRENGTH OF THE IMPROVED GROUND

Since the error in the obtained compliance ratio depends on the number of post-construction surveys of the improved ground, the number of post-construction surveys and the error in the assumed compliance ratio are summarized. Table 8.3 summarizes the relationship between the number of surveys for each compliance ratio and the errors that occur in the compliance ratio based on Figure 8.7.

Figure 8.7 shows that the post investigation case of Airport A is generally consistent with the theoretical line, and therefore, the theoretical line can be applied to the relationship between the number of investigations and the error of the compliance ratio in the improved ground. Table 8.3 is therefore similar to the theoretical line in Figure 8.7, but for an implementation injection rate of 60% or more.

As shown in Table 8.3, the smaller the conformance rate, the greater the need to increase the number of surveys. It can also be seen that the number of surveys must be increased in order to reduce the error in the fit rate. Note that the number of surveys required for statistical processing in the case of a 100% fit rate is at least 30.

When focusing on the 5% error in Figure 8.7, the number of surveys is 30 for a theoretical line with a 98% fit rate. The number of investigations required for improved ground with a conformance rate of 98% or higher is considered to be 30 or more.

As an example of how to establish the number of investigations required for improved ground at Airport A, we show how to establish the number of investigations required for improved ground. Airport A was an in-service

TABLE 8.3 Errors and Estimated Number of Surveys Required

	Number of Surveys Corresponding to Errors				
Error (%)	Rate of Injection Implemented 100~98%	Rate of Injection Implemented 97~90%	Rate of Injection Implemented 89~80%	Rate of Injection Implemented 79~70%	Rate of Injection Implemented 69~60%
5	30[a]	140	250	320	370
10	30[a]	30[a]	60	80	90
20	30[a]	30[a]	30[a]	30[a]	30[a]

Note:
[a] 30 or more as the number of surveys required for statistical processing.

runway at an airport, and the issue was to ensure the bearing capacity of the runway in the event of an earthquake. Therefore, we attempted to establish the number of investigations required for geotechnical evaluation based on the relationship between the degree of safety of bearing capacity and the compliance ratio (Kasama et al., 2022) shown in Figure 8.8.

As shown in Figure 8.8, the bearing capacity safety factor of Airport A was confirmed to be greater than 1.0 for any ground variation (coefficient of variation = 0.2 to 1.0), if the compliance ratio was generally greater than 60%. In order to evaluate the bearing capacity on the safe side, the required number of investigations was set using the results of the analysis with a coefficient of variation of 0.2, which has the lowest degree of bearing capacity safety, as shown by the red dashed line in Figure 8.8.

Thus, for example, if the compliance ratio of the improved soil is 70%, the allowable error to ensure a bearing capacity safety of 1.0 or more is 10% (compliance ratio of 70% ± error must exceed the required compliance ratio of 60%), and the number of investigations required is 80, according to Table 8.3.

If the conformity ratio is 80%, the allowable error to ensure a bearing capacity safety of 1.0 or more is 20%, and the number of surveys required

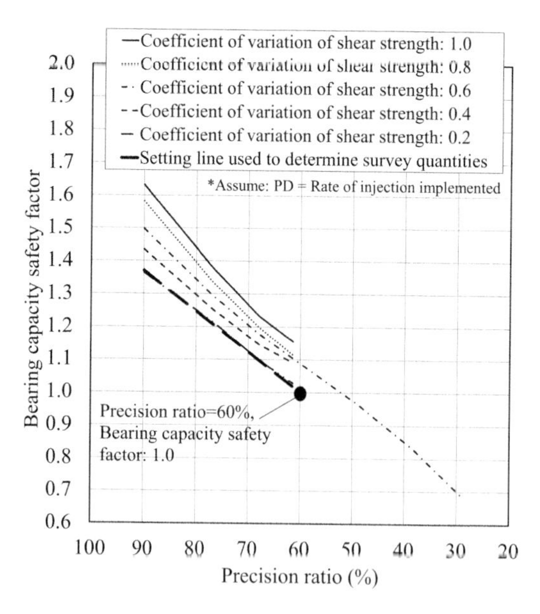

FIGURE 8.8 Relationship between bearing capacity safety factor and precision ratio.

is 15 from Figure 8.7. However, 30 surveys are required to meet the minimum number of surveys needed for statistical processing.

As in this case, an acceptable compliance ratio that satisfies the required performance (in the above example, a compliance ratio of 60% that satisfies a bearing capacity safety level of 1.0 or higher) can be set to determine the number of investigations required.

8.5 METHOD OF INVESTIGATION OF IMPROVED GROUND, EXAMPLE OF SETTING THE NUMBER OF INVESTIGATIONS

This section shows an example of setting the investigation method and number of investigations for the improved ground at Airport A based on "8.4 Number of Investigations Required to Confirm the Strength of the Improved Ground" described above. The area of improvement per block at Airport A is 60 m × 20 m, and the thickness of the improvement layer is 4 m. Therefore, the amount of soil to be improved is 4800 m^3.

The target blocks in the example setup are assumed to be three blocks, each adjacent to the other, and the injection rates for each block are assumed to be 70%, 80%, and 100%. No autocorrelation was observed in the improved ground in the subject area, and the ground was assumed to be random. The PDC test was used as the survey method because it enables continuous data acquisition (at intervals of 20 cm) in the depth direction.

As shown in "Section 8.4 Number of Investigations Required to Confirm the Strength of the Improved Ground," when focusing on the bearing capacity problem at Airport A, if the compliance ratio is 70%, an error of up to 10% is allowed, and therefore, from Table 8.3, it is necessary to obtain at least 80 investigations. Therefore, the thickness of the improvement layer per site is about 4 m. The PDC test yields about 20 surveys at 20 cm intervals, which means that the number of survey sites that need to be secured is four.

Similarly, if the fit is 80%, an error of up to 20% is allowed. However, in the case where blocks with a 70% conformance ratio are contiguous, as described above, it is suggested that the error of the block in question be set to 10%—the same as the adjacent blocks with a 70% conformance ratio—as a safe setting to ensure the overall bearing capacity of the improved ground.

In this case, it is necessary to obtain at least 60 surveys from Table 8.3. Therefore, the number of survey sites that need to be secured is three, since the number of surveys per site is about 20 as well.

Note that if the fit rate is 100%, it is necessary to obtain at least 30 surveys for statistical processing. For this reason, the number of surveys per site is similarly about 20, so the number of survey sites that need to be secured is two, and the number of surveys to be obtained is 40.

Figure 8.9 shows an example of a survey point layout based on the number of survey points for each of the above-mentioned compliance ratios.

It should be noted that the target area has been confirmed to be a random ground with no autocorrelation, and the placement of survey points

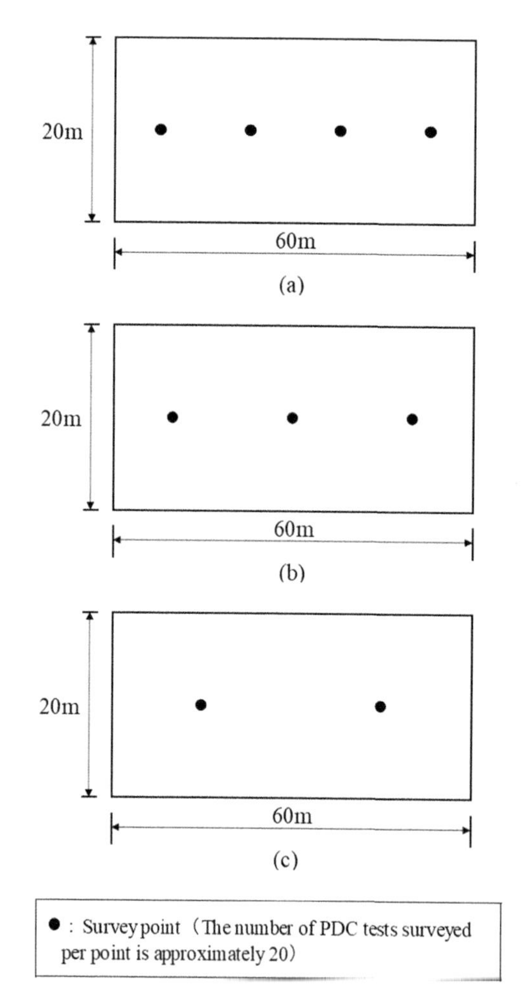

FIGURE 8.9 Example of survey sites (a) rate of injection implemented 70%: 4 points, (b) rate of injection implemented 80%: 3 points, (c) rate of injection implemented 100%: 2 points.

is not restricted if a specified number of survey points can be secured within the area of interest.

When evaluating the strength after construction in the subject area, it is necessary to confirm that the improved ground in the subject area is random ground after evaluating the autocorrelation coefficient based on the obtained post-construction survey results.

The number of survey points per block based on the above proposal would be 2 to 4 points, while the number of survey points per block based on the current Communication (CDIT, 2020) would be 6 points, which is the same level as the number of survey points based on the number of survey points of 3 points and depth of 2 points per block.

On the other hand, the number of surveys in our proposal would be 40 to 80 per block, which is significantly larger than the six surveys in the current Communication (CDIT, 2020). This is because the proposed survey is based on a theoretical evaluation of the heterogeneity of improved ground with less than 100% conformance and the number of surveys, while the Communication (CDIT, 2020) is based on past construction results targeting improved ground with 100% conformance.

In the case of a 100% conformance rate, the number of surveys per block is six according to the current Communication (CDIT, 2020), and in general, one survey is obtained for every 1.0 m of sample taken, resulting in a total target layer thickness of 6.0 m. On the other hand, in this proposal, the number of surveys per block is 30, and the target layer thickness is 6.0 m in total because the number of surveys is obtained at intervals of 20 cm in the depth direction, which is generally consistent with the current Communication (CDIT, 2020).

It should be noted that it is necessary to select a survey method that can ensure the required number of surveys from the improved strength evaluation method presented in Chapter 3. For the blocks with conformance rates of 70% and 80%, the error was standardized to 10% on the safe side because they are adjacent to each other. If these blocks are located far apart, the number of surveys could be planned with an error set for each block.

Since the above number of surveys is an example based on one of the results of the bearing capacity analysis for Airport A, an error of 10% may be focused on in the survey plan when indicators such as bearing capacity safety are not available. In the *Technical Manual for Permeation Grouting Method* (CDIT, 2020), a filling rate α of 90% is used for homogeneous sand

and gravel and sandy substrates when calculating the design injection volume of the chemical solution.

This means that the upper limit of the chemical grouting volume is 90% of the pore space of the ground under consideration, resulting in a difference of 10%. If there is no index to set the error, it is assumed that the above-mentioned filling ratio of 90% is used as an example, and the 10% difference in the design value of the chemical grouting volume that occurs when the filling ratio is not taken into account is considered as an error in the survey planning. In the future, it will be necessary to accumulate results based on this chapter and to evaluate the validity of the method used to set the number of surveys in this chapter.

8.6 CONCLUSIONS

For spatially heterogeneous improved ground, we proposed a method for setting the number of investigations based on a new perspective focusing on the compliance rate (implementation injection rate). An example of setting the number of surveys according to the conformance rate (implementation injection rate) is shown using a field board as an example. In the future, it will be necessary to accumulate results based on this chapter and to evaluate the validity of the method used to set the number of surveys in this chapter.

REFERENCES

Ang, A. and Tang, W.: Probability concepts in engineering. *Emphasis on Applications in Civil & Environmental Engineering*, 332, 2007.

Coastal Development Institute of Technology: *Coastal Technology Library No. 55, Technical Manual on the Permeation Grouting Method* (Revised Edition), 2020.

Kasama, K., Nagayama, T., Hamaguchi, N., Sugimura, Y., Fujii, T., Kaneko, T. and Zen, K.: Performance-based evaluation for the bearing capacity of ground improved by permeation grouting method. *Journal of Japan Society of Civil Engineers Proceedings C (Geotechnical Engineering)*, 78, 1, 45–59, 2022.

Matsui, K.: *Study on Bearing Capacity Evaluation of Cast in Place Friction Piles with Consideration of Uncertainty of Soil Properties*, Doctoral thesis, Kyushu University Japan, 65–72, 1992.

Ports and Harbours Bureau and Civil Aviation Bureau, Ministry of Land, Infrastructure, Transport and Tourism: Methods of Investigating Ground Improvement Effectiveness for Ground Improvement Work Using Chemical Injection Methods, etc. (Attachment), 1–6, 2017.

Statistics Section, Department of Social Sciences, College of Arts and Sciences, The University of Tokyo: Basic Statistics I Introduction to Statistics, University of Tokyo Press, 202, 1991.

Sugano, T., Zen, K., Suemasa, N., Kasugai, Y., Yamazaki, H., Hayashi, K., Sawada, S., Endo, T., Kato, T., Nakagawa, H., Kiku, H., Yamaguchi, E., Fujii, N., Baba, K., Fujii, T. and Takada, K.: Study on strength evaluation technique by using in-situ tests on chemical grouted ground as a countermeasure for liquefaction, Technical Note of the Port and Airport Research Institute, No. 1366, 2020.

Index

Pages in *italics* refer to figures and pages in **bold** refer to tables.

For Product Safety Concerns and Information please contact our EU representative GPSR@taylorandfrancis.com Taylor & Francis Verlag GmbH, Kaufingerstraße 24, 80331 München, Germany